Comparative Population History of Eastern Asia

Toru Suzuki

Comparative Population History of Eastern Asia

Springer

Toru Suzuki
Matsudo, Chiba, Japan

ISBN 978-981-99-9366-6 ISBN 978-981-99-9367-3 (eBook)
https://doi.org/10.1007/978-981-99-9367-3

This Springer imprint is published by the registered company Springer Nature Singapore Pte Ltd.
The registered company address is: 152 Beach Road, #21-01/04 Gateway East, Singapore 189721, Singapore

Paper in this product is recyclable.

Preface

The idea of this book gradually developed while I worked at the National Institute of Population and Social Security Research (IPSS), Tokyo, Japan. I published two books from a series of Springer Brief on Population Studies of Japan based on outcomes from research projects supported by Health and Labour Sciences Research Grants (Ministry of Health, Labour and Welfare, Government of Japan). In my first book, *Low Fertility and Population Aging in Japan and Eastern Asia* (2013), I limited my interest to fertility decline in Eastern Asia since 1980. However, I realized that a longer historical view was required to study the contemporary population issues in this region. I discussed historical roots of early aging in China, compressed urbanization in Korea, and cross-border migrations in Eastern Asia in addition to extremely low fertility in my second book, *Eastern Asian Population History and Contemporary Population Issues* (2019). However, the book focused on modern and contemporary history and the premodern era was discussed only briefly.

This book aims at an extensive comparative population history of Japan, Korea, China, and Taiwan starting from the origin of each ethnic group. In addition to highlighting different demographic trajectories, this book examines demographic impacts of interactions such as wars, migrations, and transmission of infectious diseases. One purpose is to review works on historical demography in each country written in local language and introduce in English. After I retired from the IPSS in 2020, I could stay at the Seoul National University as a guest professor and collected Korean literatures. I gave up my plan to stay in Taiwan or China to collect Chinese literatures due to the COVID-19 pandemic. Thus, I'm afraid of missing some important works in Chinese.

I would like to thank my colleagues in IPSS and outside participants of research projects on Eastern Asia where I worked as the principal investigator. With the support of the scientific research grants, I could discuss with many Eastern Asian scholars, researchers, and government officials. Although there are too many names to list here, I'm especially grateful to Prof. Kojima Hiroshi, Prof. Ito Shoichi,

Prof. Cho Young-Tae, Prof. Park Kyeong-Suk, Prof. Jun Kwang-Hee, Prof. Kim Doo-Sub, Prof. Chen Yu-Hua, Prof. James C. T. Hsueh, Prof. Li Meilin, Dr. Yang Wen-Shan, Dr. Lin Ji-Ping, and Dr. Yu Rouh-Rong.

Chiba, Japan Toru Suzuki

Contents

Chapter 1
Introduction—Present and Future of Eastern Asia

Eastern Asian societies have been the frontrunner both in economic and demographic changes. Japan used to be the only advanced country in this region and became the second largest economy worldwide in the 1970s. At that time, the possibility of "Japan as number one" was seriously discussed (Vogel 1979). Until the 1970s, it was believed that no developed country other than Japan would appear from Asia or Africa. However, the Republic of Korea ("Korea," hereinafter), the Republic of China ("Taiwan," hereinafter), Hong Kong, and Singapore achieved impressive economic success since the 1970s and were called "four dragons." Korea started from an extremely underdeveloped condition after the Korean War to become a noticeable economic power through the "miracle on the Han River." Although Taiwan does not have a heavy industry and produces no automobile, train, or ship, it achieved a great success only by light and hi-tech industries. The People's Republic of China ("China," hereinafter) was slower to start than other Eastern Asian newly industrialized economies (NIEs) because of improper economic policies led by the Communist party. However, after the economic reform in the 1980s, the development of China accelerated and replaced Japan as the second largest economy in 2011.

In the course of economic development, Eastern Asian contributions to science, technology, arts, and culture have been increasing. Japan has won many novel prizes on natural sciences and has materialized various inventions including high speed rail, hybrid vehicle, video game console, instant noodles, etc. Japanese conventional culture is popular around the world and haiku, conventional short poem, has been transmitted into various languages. The pop culture is also competitive and numerous works on movies, TV dramas, animations, novels, and manga are circulating worldwide. The Korean wave since the 2000s was epoch-making in the sense that Korea has become a net exporter of cultural contents. Chinese traditional cultures including classic philosophy, poems, novels, theatrical dramas, and martial arts, have been influential. In accordance with China's economic growth, a huge number of foreigners are learning Chinese language, culture, and history. Thus, China's influence will continue growing in the future.

T. Suzuki, *Comparative Population History of Eastern Asia*,
https://doi.org/10.1007/978-981-99-9367-3_1

Demographically, Eastern Asia is characterized by the lowest level of fertility and the anticipated highest percentage of elderly persons. A total fertility rate (TFR) of less than 1.0 was observed only in Taiwan and Korea except for small area. An incredibly low TFR of 0.895 was recorded in Taiwan in 2010 and 0.84 was recorded in Korea in 2020. Many countries in Southern and Eastern Europe and in the former Soviet Union experienced lowest-low fertility, defined as TFR is 1.3 or less (Kohler et al. 2002). While such low fertility lasted for less than twelve years in those European countries, TFRs in Taiwan and Korea have been less than 1.3 for more than twenty years. It is ambiguous when Taiwan and Korea can escape lowest-low fertility. The United Nations Population Division (UNPD) assumed in its medium variant that Taiwan will regain the TFR of 1.41 in 2050 but the TFR of Korea will stay at 1.17 (UNPD 2022). However, the official population projection in Taiwan is more pessimistic. The medium variant of National Development Council (NDC 2022) of Taiwan assumed that the TFR will stay at 1.20 in 2050, while the medium variant of Statistics Korea (2021) assumed that the TFR of 2050 will be 1.21. If these are correct, lowest-low fertility of not only Korea but also Taiwan will last for more than half a century.

Extremely low fertility brings about extremely acute population aging. Presently Japan is the most aged country in the world. This position owes to lower mortality since the late 1970s in addition to lower fertility during the late 1950s and 1960s than other developed countries (Suzuki 2013). It is certain that Japan will be overridden by Taiwan and Korea in the predictable future. The UNPD (2022) expected that the percentage of the elderly aged 65 years and older in Korea will excess that of Japan by 2050. The projection also showed that Taiwan will override Japan by 2060. The timing will be earlier if pessimistic assumptions in official projections of Korea and Taiwan are correct. In any case, three most aged societies will concentrate in Eastern Asia in the near future. According to the UNPD, the percentage of the elderly in China will not exceed that of Japan by 2070. The trajectory of population aging is less certain than other Eastern Asian countries because of unreliable fertility statistics in China. However, the population aging in China can cause more serious problems because it proceeds at a lower developmental level than Japan, Korea, or Taiwan.

Confucian societies such as China, Taiwan, and Korea are characterized with the strong son preference and have experienced the abnormally high sex ratio at birth. In this sense, Japan is not a Confucian society because there was no such distortion. The sex ratio at birth in Confucian societies started rising since the 1980s when the sex identification technology enabled selective abortion. While the ratio returned to the normal level in Korea around 2010, it is still abnormally high in China and Taiwan. It is worried that the distorted sex ratio can cause a serious marriage squeeze of men and human trafficking of single women in China. It is also possible that a huge number of single men will get older without any familial support and will be burdensome for the national welfare system.

Japan has been the top runner in mortality decline. Recently, however, Japan was overridden by Hong Kong in terms of the life expectancy at birth. It is possible that Korea and Taiwan catch up Japan in the near future. Such low mortality can accelerate the population aging in Eastern Asia. The process of mortality decline in

Eastern Asia is an important issue in this book. In addition to wars in the 1940s and 1950s, disastrous famines and pandemics are not events in the distant past.

Nuptiality decline, or the trend of late and less marriages, has a critical impact on extremely low fertility in Eastern Asia. Socioeconomic changes such as female labor force participation, rising cost of childcare and education, worsened labor market for young workers, and increasing uncertainty about the future have been blocking the decision to marry and have a child. Since extramarital births are rare in Eastern Asia, a decline in marriage directly links to fertility decline. Why cohabitations and extramarital births do not increase as in Europe or Northern America will be discussed in Chap. 6. The divorce rate in Eastern Asia is higher than many Southern and Western European countries. Although Confucian societies such as China, Taiwan, and Korea used to show lower divorce rates than Japan, the situation has been reversed since the 2000s.

The population distribution changed drastically under the modernization since the middle or late nineteenth century. While the impact of urbanization appeared in Japan as early as in the 1920s, it was in the 1950s that the concentration of population to urban areas became evident in Korea, Taiwan, and China. Acute urbanization and concentration in the capital area have caused various problems, including degradation of living conditions, rise in living costs, and environmental destruction. The urbanization of Korea has been more compressed than in Japan or Taiwan. The percentage of population living in the capital area (Seoul, Incheon, and Gyeonggi province) increased from 18.3% in 1955 to 49.1% in 2010. President Roh Moo-Hyun proposed the relocation of the capital to Chungcheong province, and it was agreed that while a part of the government remains in Seoul, other governmental offices move to the newly built special administrative city of Sejong. Although relocation was mostly completed by 2015, its impact on population distribution is evaluated to be small.

Throughout the modern and contemporary eras, Koreans have been the most mobile people in Eastern Asia. The differential mobility between Koreans and the Taiwanese derived from the different agricultural development under the Japanese rule. Because cities and frontiers in Korea could not absorb the rural surplus population, a massive outflow headed to Manchuria, China, Japan, and Sakhalin. The exodus of surplus population from Korean villages continued after the Korean War to induce the compressed urbanization and the increase in overseas Koreans. Although the absolute number of overseas Chinese is enormous, the ratio to the home population is lower than that of overseas Koreans.

Japan has been slow in accepting foreign laborers. While Korea and Taiwan launched the Employment Permission System in 1992 and 2003, respectively, Japan accepted manual laborers only through the industrial trainee program. As a result, the percentage of foreign population in Japan was lower than in Korea or Taiwan. However, a labor shortage due to the declining population forced the Japanese government to open its labor market further. Competition between Japan, Korea, and Taiwan to invite productive foreign laborers may accelerate in the future. The feminization of cross-border migration became prominent in Eastern Asia after the turn of the century. At its peak, cross-border marriages in Taiwan, mainly between Taiwanese grooms and Chinese brides, accounted for more than 30% of total marriages. An unexpectedly rapid increase in the foreign population may cause various problems, including the labor exploitation and human rights issues. The position of China in cross-border migration is double-sided. While China sends laborers and brides to Japan, Korea, and Taiwan, there are increasing demands to accept migrant workers and brides mainly from Southeastern Asia.

References

Kohler, Hans-Peter, Francesco C. Billari, and José Antonio Ortega. 2002. The emergence of lowest-low fertility in Europe during the 1990s. *Population and Development Review* 28 (4): 641–680. https://www.doi.org/10.1111/padr.2002.28.issue-4, https://www.doi.org/10.1111/j.1728-4457.2002.00641.x

National Development Council, Republic of China. 2022. *Zhonghua Mínguo Renkou Tuigu* (in Chinese).

Statistics Korea. 2021. *Jangrae Ingu Chugye: 2020–2070-nyeon* (in Korean).

Suzuki, Toru. 2013. *Low Fertility and Population Aging in Japan and Eastern Asia*. Springer.

United Nations Population Division. 2022. *World Population Prospects, the 2022 Revision.*

Vogel, Ezra F. 1979. *Japan as Number One: Lessons for America*. Harvard University Press.

Chapter 2
Origins of Eastern Asian Peoples

This book discusses populations of Japan, Korea, China, and Taiwan. After the division of Korea in 1946, the Republic of Korea in the south has been more important both economically and demographically. Thus, demographic changes of the Republic of Korea will be compared with Japan, China, and Taiwan. The population of the Democratic People's Republic of Korea (DPR Korea) in the north will be only briefly discussed. Although the population of Mongolia after the modern period is not discussed in this book, the Mongols were extremely influential in the premodern history of Eastern Asia. Thus, the origin of Mongolic and other Altaic language groups will be discussed in this chapter.

While a population includes both natives and foreigners, the size of each is apparent today because Eastern Asian nations do not allow dual nationality. In Eastern Asia, the percentage of foreign population is less than 5%. The native population is also homogeneous so that the largest ethnic group accounts for more than 90%. The superiority of the Yamato Japanese in Japan, Han Koreans in Korea, the Han Chinese in China continued from the ancient time. Linguistically, the Han Chinese can be divided into dialect groups such as Mandarin, Wu, Min, Yue, and Hakka. Although 99% of residents in Taiwan are the Han Chinese, the society is considered as multi-ethnic with this detailed classification. According to the 2020 census of Taiwan, 65.9% used Minnan as the first language, followed by Mandarin (28.5%), Hakka (4.8%), and indigenous languages (0.6%). The majority group of Taiwan using Minnan, a variant of Min Chinese, are called Hoklo.

The origin of an ethnic group can be studied through genetics, archaeology, linguistics in addition to historical studies (Sagart et al. 2005). The recent development in the molecular biology, especially that in the PCR test of ancient bones, brought about various new findings on migration and diffusion of Homo sapiens since the exodus from Africa in 60,000 BP. Archaeological analyses of architectures, crafts, and other remains also provides important clues on distribution and migration of various ethnic groups. Since the ethnicity is a sociocultural concept, the language is more important than genes and archaeological remains to identify an ethnic group.

T. Suzuki, *Comparative Population History of Eastern Asia*,
https://doi.org/10.1007/978-981-99-9367-3_2

For example, although the Jomon people left numerous remains and their DNA is preserved by the Japanese today, they were certainly not the speakers of the proto-Japanese language (Diamond 1998; Robbeets et al. 2021) and cannot be called "the Japanese." This book considers that the establishment of the ethic Japanese is equivalent with the appearance of the proto-Japanese language and discusses the process of migration and diffusion of proto-Japanese speakers. Origins of ethnic Koreans, ethnic Chinese, and other ethnic groups in Eastern Asia are also examined through the developments and expansions of language groups.

The study of language families started from the discovery of the Indo-European language family. Since an evident phonological correspondence was found in this family, such a sound law was considered as the precondition to define a language family. In Eastern and Southeastern Asia, many language families including Sino-Tibetan, Austroasiatic, and Austronesian were identified. In the case of Altaic languages, however, it was considered that phonological correspondences between Turkic, Mongolian, and Tungusic languages were too weak to define the "Altaic language family." Although the Japanese and Korean languages share various common aspects with Altaic languages, sound correspondences are even weaker and could not related with any other languages (Kudo 2005; Tanaka 2021).

More recently, various superfamilies, or phyla, were proposed being based on syntax and lexicology in addition to phonology (Manning 2006). In this situation, Eastern Asian languages belong to one of three superfamilies: Dene-Caucasian, Austric, or Eurasiatic. The Dene-Caucasian superfamily is a provocative hypothesis that Sino-Tibetan languages derived from a common ancestor of the ancient Sumerian language, the Basque language at the France-Spain border, and North Caucasian languages. The Austric superfamily includes Austroasiatic, Austronesian, Tai-Kadai, and Hmong-Mien languages that are widely spoken in Southeastern Asia, China, and Taiwan. The Eurasiatic superfamily hypothesis asserts that the Japanese, Korean, and Altaic languages have a common ancestor with Indo-European, Ularic, and Eskimo-Aleut languages. The next section discusses migration of Homo sapiens after the exodus from Africa and diffusion into Eastern Asia.

2.1 Diffusion of *Homo sapiens* to Eastern Asia

All the ethnic groups, or language families, originated in Africa as far as they are Homo sapiens. While Homo sapiens attempted the exodus from Africa several times, emigration that resulted in today's distribution took place around 60,000 BP and the arrival at Eastern Asia was around 47,000 BP (Shinoda 2020). Manning (2006) distinguished two stages of human migration, namely the primary diffusion in tropical regions and the secondary penetration into temperate regions. Figure 2.1 summarizes his discussion on the differentiation of language groups. After the exodus from Africa, the Dravidan, the Indo-Pacific, the Australian, the Dene-Caucasian, and the Austlic superfamilies appeared in tropical regions. The first group that migrated into tempered region about 40,000 BP was the Dene-Caucasian that moved

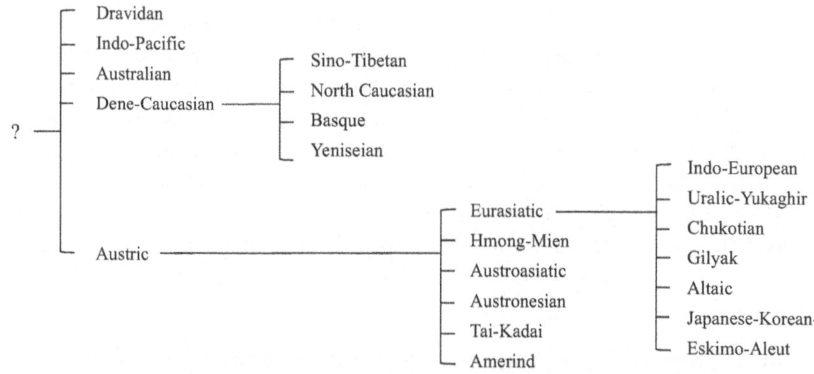

Fig. 2.1 Language superfamilies. Based on Manning (2006)

from Yunnan province to Sichuan, Chinghai, Central Asia, Caucasus, and Europe. About 35,000 years ago, a group of fishers, probably Austlic speakers, moved from Southeast Asia to northwest Asian coast. Manning assumed that Eurasiatic and Amerind superfamilies emerged in this center of linguistic diversity including Chukotian, Gilyak, Japanese-Korean-Ainu, and Altaic languages. According to him, Indo-European and Uralic language families sprang from groups moved westward from this region, while ancestors of native Americans moved eastward. As the Eurasiatic superfamily occupied temperate Eurasia, offspring of the first wave migrants other than Sino-Tibetan speakers were isolated as minority groups. In Southeast Asia, Austroasiatic, Austronesian, Tai-Kadai, and Hmong-Mien language families derived from the Austlic speakers remained in this region.

According to Manning, Eurasiatic superfamily emerged in the western shores of the northern Pacific and spread westward across the Eurasian steppes during the Paleolithic period. While it is unknown if Eurasiatic languages were already dominant in temperate Eurasia before the Neolithic Revolution, the transition to agrarian societies certainly caused a drastic change in the linguistic map of the Eurasian continent. The Indo-European language family is one of the most successfully dispersed groups consequent upon the establishment of agricultural economy (Bellwood 2005). These languages contributed to the foundation of comparative linguistics and their relatedness has been studied more extensively than other groups. The Indo-European group has an advantage that there remain many ancient documents in Sanskrit, Avestan, Mycenaean Greek in addition to Hittite language that goes back to the sixteenth century BC. Its relatedness is well established and genetic features proper to its speakers has been investigated intensively. Even for Indo-European languages, however, it is difficult to prove its origin by archaeological evidence. While the Kurgan hypothesis asserted that the proto-Indo-European speakers were nomadic pastoralists came from the Dnieper–Volga region, the Anatolian hypothesis assumed that they were farmers originated in Neolithic Anatolia (Kazama 1993).

Eastern Asian languages are less advantageous than the Indo-European languages in exploring their origins. No language other than Old Chinese had written records in the Bronze Age. Own writing system was established in the eighth century for Japanese, the thirteenth century for Mongolian, the fifteenth century for Korean, and the sixteenth century for Manchu language. Although some words were recorded by the ancient Chinese, they are insufficient to reconstruct proto-languages. The lack of ancient literatures must be one reason why sound laws for Altaic languages cannot be established. The slow development of own writing system also limits historical studies. The population distributions and migration processes of each ethnic group can be investigated only through fragmental records in ancient China.

2.2 Sino-Tibetan Language Family and Formation of Han Chinese

According to Manning's scenario, the Sino-Tibetan language family derived from the first wave migrants from tropical to temperate regions 45,000 BP. While the proto-Sino-Tibetan language appeared 7000–6000 BP, its homeland is under controversy. Candidates include northern India, Sichuan Province, the Tibetan Highlands, and the Yellow River basin (Sagart et al. 2005). According to van Driem (2005), the Tibeto-Burman family originated in Sichuan and the Sino-Bodic language derived from its Northeastern group. He asserted that the Sintic language group migrated from Sichuan to the Yellow River basin before the beginning of the sixth millennium BC. On the other hand, Starosta (2005) discussed that the pre-Sino-Tibetan, the proto-Hmong-Mien, the proto-Austro-Asiatic, and the proto-Austronesian language groups originated from the proto-East Asian group inhabited in the North China Plain between the Yellow River and the Yangzi River in 6500–6000 BC. This implies that the ancestors of the Han Chinese already settled in the Yellow River region in the early Neolithic period.

China was isolated from other cradles of civilization. While Egypt and India accepted agriculture and pastoralism from Western Asia, the Neolithic Revolution in China took place independently from other cradles. The Chinese domesticated rice, foxtail millet, sorghum bicolor that were different from Western crops such as wheat, pea, and lentil. Pig was also originally domesticated in China (Diamond 1997). On the other hand, the writing system, horse riding and horse-drawn vehicles, the use of metal were possibly transmitted from the west to China. Oracle bone script, which is the oldest remaining character, appeared in the Shang dynasty established around 1700–1600 BC. Although the genealogical relation is unknown, it is plausible that the ancient Chinese borrowed the idea of writing system from the west where scripts have older history than China. Cuneiform in Mesopotamia goes back 3200 BC, Hieroglyph in Egypt appeared around 3000 BC, and the Indus scripts was used around 2600 BC.

It is presumed that horse was domesticated in the Black Sea coast around 6000 BP. According to the Kurgan hypothesis, a branch of the Indo-European speakers moved from the lower Volga and Kazakhstan to the Black Sea region with domesticated horses. It was in the Longshan period (3000–1900 BC) that horses were domesticated in China, probably under the western influence. While wheeled vehicles were used since 2500 BC and developed into a chariot during the Bronze Age in Europe (Kazama 1993), a chariot in China appeared in the Shang dynasty (Okuno 2000).

Bronze was used since 3500 BC in Western Asia, 2300 BC in Europe, and 2100 BC in China. Bronze artefacts in the Lower Xiajiadian culture (2200–1600 BC) were under the influence of the Andronovo culture in southern Siberia, suggesting that there was a route of transmission from the west to the east (Hayashi 2017). While a small amount of artefacts using iron meteorite were already used during the Bronze Age, mass production of iron tools started in Anatolia in 1500 BC. It was in the late Western Zhou period (ca 770 BC) that iron farm tools spread in China (Okuno 2000). Thus, it is supposed that the main flows of peoples and cultures in the ancient Eurasian continent were those from the west to the east.

While China accepted various cultural elements from the western cradles, it is the only ancient civilization that the ethnic identity continues until today. Oracle bone script in the Shang dynasty transformed into the Chinese characters used today. Although China was conquered several times by Altaic nomads, the Han Chinese never switched to conqueror's languages. Confucianism established in the fifth century BC continues to be the principle of Chinese life. On the contrary, Mesopotamia and Egypt were conquered by Islam in the seventh century and most people speak Arabic and believe Islam today. Neither cuneiform nor hieroglyph has direct relation with the Arabic alphabet. Zoroastrianism and other ancient religions have either declined to be minor religions or disappeared. Little is known about the language and religion of the Indus Valley civilization before the arrival of Aryan people in the fifteenth century BC. It is only in China that the complete continuity of language and religion from the cradle of civilization to today can be observed.

In the Yellow River basin, millet, barley, and bean were planted since 9500 BP. Although the wet rice cultivation in the Changjiang basin started about the same time or even earlier, no systematic writing has been discovered in this area. The contrast between the Yellow River basin in the north and the Changjiang basin in the south had important meaning in the political economy in China throughout its history. In the ancient period until the Tang dynasty (618–907), the south had less population than the north due to the unhealthy environment and underdevelopment. As a result, southern China was dominated both politically and culturally by kingdoms and dynasties in the north. Although powerful kingdoms appeared from the Changjiang basin in the Spring and Autumn period (770–476 BC) and the Warring States period (403–221 BC), the Qin dynasty in the north finally unified whole China in 221 BC. During the Qin (221–206 BC) and the Han (202 BC–220 AD) periods, the north and the south were integrated into a homogeneous ethnic group of Han Chinese. Sinocentrism that only the Han Chinese holds the genuine civilization played an important role in the cultural integration. Although the Han Chinese contains a considerable variation in languages and genes, those who respect Confucian values, practice ceremonies, and learn classics were presumed to be the Han Chinese. Racial and ethnic origins were not important and linguistic differences were covered by the written classic Chinese. The domain of Han Chinese expanded in the Han period southward to Guangdong province and northern Vietnam.

2.3 Ethnic Composition of Taiwan

Although the Han Chinese is assumed as one ethnic group speaking the same language in the People's Republic of China, the difference between dialects is larger than that between different European languages. Major dialects of Chinese include the Mandarin spoken in northern and western China, the Wu spoken in Zhejiang and Jiangsu, the Min spoken in Fujian, the Yue spoken in Guangdong and Guangxi, and languages of Hakka people scattered in southern China. In Taiwan, these dialects are supposed to be the marker of different ethnic groups. Four major ethnic groups in Taiwan are the Hoklo people speaking the Min, the Mainlanders speaking the Mandarin, the Hakka people, and the indigenous Taiwanese speaking Austronesian languages.

Linguists agree that Taiwan was the homeland of the Austronesian language family that spreads in the Pacific and Indian Oceans, from Madagascar to Easter Island, and from Hawaii to New Zealand. According to Starosta (2005), one group of pre-Austronesians crossed the Taiwan Strait from Fujian and landed on the southwest coast of Taiwan. Those who remained in Fujian were absorbed by the successive waves of Hmong-Mien, Tai-Kadai, and Sino-Tibetan peoples and eventually disappeared. In Taiwan, those who inhabited the east coast developed sophisticated seafaring techniques. One group landed on the northern Philippines became the ancestors of Malayo-Polynesians and eventually populated most of the Pacific land areas except for Papua-New Guinea and Australia.

Until the sixteenth century, only a small number of the Chinese settled in Taiwan. In the early seventeenth century, the Dutch and the Spanish competed in colonizing Taiwan. The Dutch drove out the rival in 1642 and encouraged Chinese farmers to migrate from Fujian and civilize western Taiwan. During the turmoil of Ming-Qing transition, a Ming general Koxinga landed Taiwan with his army and expelled the Dutch in 1662. After 21-year rule by his son and grandson, Taiwan was eventually annexed to the Qing dynasty in 1683. Although the Qing banned migration to Taiwan, the flow of Hoklo people did not stop. Hakka migration from Guangdong started as seasonal agricultural workers. Afterward, the population pressure in Guangdong promoted permanent migration (Ho 1959).

After the period of Japanese rule (1895–1945), Taiwan was requisitioned by the Republic of China. The nationalist government led by Chiang Kai-shek (1887–1975) was challenged by the communist forces led by Mao Zedong (1893–1976) immediately after the surrender of Japan. When the Chinese Civil War eventually ended with the victory of the communists, one million nationalists retreated to Taiwan in 1949. The Mainlanders in Taiwan are the offspring of those nationalists that monopolized the political power until the democratization in the late 1980s.

2.4 Altaic Language Family and Japanese-Korean-Ainu Group

Greenberg (2000) included the Altaic and the Japanese-Korean-Ainu languages into the Eurasiatic superfamily together with Indo-European, Uralic-Yukaghir, Eskimo-Aleut, Chukotian, and Gilyak languages. The Altaic language family consists of the Turkic, Mongolic and Tungusic languages but its position as a language family is still controversial due to the lack of the sound law. However, these languages share impressively common phonological and grammatical features such as the avoidance of "R" at the beginning of word, the existence of vowel harmony, the lack of article or gender, the agglutinative suffixing morphology, and so forth (Tanaka 2021).

Lee (2020) asserted that the Liao River Civilization was established by the Altaic speakers in 6200 BC and that both Koreans and the Japanese derived from this "Liao River ghost population." According to him, those Altaic speakers occupied the Korean peninsula in 3000 BC and arrived at Japan in the fifth century. Probably he misunderstood Egami's nomadic conqueror hypothesis (Egami 1967) as the arrival of the ethnic Japanese. In fact, the Japanese language was already spoken in western Japan in the third century (Osada 2010).

Robbeets and her coauthors also concluded that the common ancestors of the Transeurasian family including Turkic, Mongolic, Tungusic, Korean and Japanese languages were farmers cultivating millets in the West Liao River region about 9000 BP (Robbeets et al. 2021). Those proto-Transeurasian farmers started planting rice and wheat in the early Bronze Age and transmitted those crops to the Korean peninsula about 3300–2800 BP and to Japan after 3000 BP. Genetically, the proto-Transeurasian speakers were very much different from the Jomon people inhabited in Japan and Korea but close to the proto-Sino-Tibetan speakers in the Yellow River basin.

Both Lee and Robbeets assumed that the proto-Koreans and the proto-Japanese arrived after they acquired the wet rice planting. However, it is thought that rice was transmitted from the Shandong peninsula to Korea and Japan (Fujio 2015). In that case, it is likely that those migrants that transmitted rice were the Sino-Tibetan speakers. Since the indigenous Jomon people did not prefer wetlands in lower reaches of a river, those Sino-Tibetan farmers could easily occupy lands appropriate to rice paddies. Because the rice planting enabled a higher population growth rate, it is likely that the Jomon languages and genomes in southern Korea and western Japan were replaced gradually by those from Shandong. Then, the arrival of the proto-Koreans or the proto-Japanese was much later than the timing proposed by Lee or Robbeets.

2.5 Jomon People and Ainu Language

It was 40,000–35,000 BP that Homo sapiens landed on the Japanese archipelago (Harari 2011; Saito 2017). After the case of Paleolithic hoax in 2000, the existence of Homo other than sapiens in Japan was refuted. Earthenware potteries with so-called the Jomon style appeared 16,000 BP, which marks the beginning of the Neolithic Period of Japan. The Jomon people started a settled pattern of living with a small-scale agriculture in addition to hunter and gathering. According to Kito (2007), the population reached about 260,000 in 2300 BC. The population concentrated in eastern Japan where fish catches and nuts were richer than western Japan. Kito estimated that the Jomon population declined to about 76,000 in BC 900 due to the climate cooling and the transmission of infectious diseases from the continent. On the other hand, Saito (2017) asserted based on a genetic analysis that the population never fell below 300,000.

The Ainu language can be a descendant of ancient languages spoken by the Jomon people. When the Yayoi people landed Japan and started wet rice cultivation 2900 BP, the Jomon people were living with hunter-gathering and fire-fallow cultivation of millets (Fujio 2015). Both "Jomon" and "Yayoi" were named after the style of pottery discovered by archaeologists. The biological differences between the Jomon and the Yayoi have been clarified by geneticists. According to a principal component analysis of Single Nucleotide Polymorphism, Jomon genomes remain in the Ainu in Hokkaido and the Ryukyuan in the Ryukyu islands (Saito 2017). This implies that the Jomon people were forced out to the northern and southern tips of the Japanese archipelago by the expansion of the Yayoi people.

It is unknown when the direct ancestors of the Ainu arrived at Japan. It is possible that the proto-Ainu speakers were already inhabited in Japan when the Jomon pottery was created in 16,000 BP. It is also possible that the Jomom people came to Japan slightly before the arrival of the Yayoi. The Jomon bones analyzed by Saito (2017) were those of around 3000 BP. If those people transmitted both gene and language, it is unlikely that the proto-Ainu speakers arrived at Japan after that period.

2.6 Yayoi People and Proto-Japanese Language

The most important difference between the Jomon and Yayoi cultures was not the rice planting itself but the introduction of wet paddy for rice. The Jomon people already utilized rice under the fire-fallow cultivation together with millet and bean. According to Sato (2002), it is likely that the slash-and-burn rice planting was transmitted from Taiwan and Okinawa to southern Kyushu. However, the arrival of rice was not as innovative as the introduction of irrigated paddy farming by the Yayoi people. Since the Japanese archipelago is characterized with rich ecological resources both from the land and the sea, the addition of rice made only a minor contribution to the Jomon society (Diamond 1998).

The Yayoi people with the wet rice cultivation came to northern Kyushu 2900 BP (Kimura 2010; Fujio 2015; Shinoda 2020). Saito (2017) discussed that either the early Yayoi people came from Korea in 2900 BP or fishers already settled in Japan at that time were the proto-Japanese speakers. However, it is more likely that the early Yayoi people were Sino-Tibetan speakers, and the Jomon and other indigenous cultures were wiped out as the Yayoi culture spread in western Japan. In northern Kyushu, indices of agrarian society such as moated settlements, large scale tombs, and warfare in addition to irrigated rice paddy appeared in the ninth century BC. However, such agrarian societies were limited in northern Kyushu and did not spread to other parts of Japan for 250 years (Fujino 2015). This implies that migration from Korean peninsula was moderate, and the population growth rate of the Yayoi people was low. Even after the seventh century BC when the Yayoi culture started to spread in western Japan, the Yayoi people settled in downstream areas where the Jomon people did not prefer. Thus, the replacement of culture did not take place for centuries.

Egami (1967) asserted that a group of Tungusic horse riders conquered western Japan and established the Yamato kingdom in the early fourth century. Although he emphasized that his hypothesis was not about the foundation of the ethnic Japanese but that of the political unification of western Japan, Lee (2020) seems to misunderstand it. In any case, Egami's hypothesis was not accepted by the Japanese historians because there was no major cultural disruption in the early fourth century. A specific style of tombs called "zenpo-koen-fun" continued between the third and the seventh century, and the language spoken in the early third century can be seen as the direct ancestor of the Japanese.

Osada (2010) wrote that the proto-Japanese inhabited southern Manchuria and northern Korea before coming to Japan. However, he did not refer to the timing of landing and the reason why language and culture could dominate in western Japan. One major cultural disruption during the Yayoi period that lasted about 12,000 years until the third century was the introduction of metalware. In Japan, bronze ware began to be manufactured in northern Kyushu in the second century BC, and ironware was also spread at that time (Fujio 2015). Thus, it can be inferred that the proto-Japanese people holding those metal weapons conquered western Japan and the early Yayoi people speaking Sino-Tibetan languages transferred to the ruler's language.

2.7 Old Korean Languages and Foundation of Han Koreans

The wet rice cultivation in Korea began 3000 BP, which was slightly earlier than that in Japan (Fujio 2015). It is likely that Sino-Tibetan migrants from Shandong found wet riverside in southern tip of the Korean peninsula and spread in today's Gyeongsang, Jeolla, and Chungcheong provinces. Those Sino-Tibetan people, as the early Yayoi in Japan, did not have metalware. An influential hypothesis about the origin of Koreans is that migrants holding bronze ware came from Manchuria and dominated indigenous people (Pang 2018). According to historical records by the Chinese, the Korean peninsula was occupied by various Altaic groups such as

Goguryeo, Okjeo, Ye-Maek, Han Koreans, and Wa Japanese in the third century. Because the peninsula is not as isolated as the Japanese archipelago, there must have been a continuous flow of Sino-Tibetan speakers.

According to *Shiji* (Records of the Grand Historian) written by Sima Qian in the first century BC, a general of Yan kingdom occupied Beijing area named Wiman fled to north-western Korea with a thousand followers in 194 BC. Wiman once served to Korean king Jun but later carried out a coup d'état and founded his own kingdom in Pyongyang area. He ruled native Koreans and Chinese refugees from the Qin dynasty that fell in 206 BC. In 109 BC, the seventh emperor Wu of Han sent an expeditionary force and killed the Korean king Ugeo, a grandson of Wiman. The Han dynasty established four commanderies to rule northern part of the peninsula. Three of them disappeared within a few decades but Lelang commandery in Pyongyang area lasted until the beginning of the fourth century. Daifang commandery was established in the beginning of the third century as the Chinese colony expanded southward to Seoul area. Both Lelang and Daifang commanderies were demolished by Goguryeo in 314. The "Treatise on Geography" in the *Book of Han* recorded that the population of Lelang commandery in 45 BC was 406,748. A large part of them must have been of Chinese ethnicity. Cultural influence of China was extensive such as the writing system, metalware, architecture in addition to various institutions for politics, economy, and military affairs. Not only various Altaic groups in the peninsula but also local groups in Japan sent missionaries to Lelang and Daifang commanderies for legitimation.

Among non-Chinese peoples in Korea, Han Koreans occupied the southern part of the peninsula in the early third century. While a branch called Mahan occupied the southeastern part (today's Chungcheong and Jeolla provinces), Byeonhan and Jinhan inhabited in the southeastern part (today's Gyeongsang province). According to an observation by the Chinese, the language of Mahan was different from that of Byeonhan and Jinhan. The Ye-Maek people mainly lived in today's Gangwon province in the east, and the Okjeo lived in today's Hamgyong province in the northeast. Japan or the Japanese was called "Wa" in China and Korea until the seventh century. A considerable number of the Wa Japanese lived in the southern tip of the peninsula and transported between Japan. Goguryeo occupied southern Manchuria, and Buyeo dominated the north of Goguryeo.

According to historical records in China, languages of Ye-Maek, Okjeo, Goguryeo, and Buyeo were affiliated, and a hypothetical Buyeo language family was proposed by modern linguists (Choi 2004). However, the Korean language derived not from the Buyeo language family but from Han Korean language, especially that of Jinhan. The kingdom of Silla appeared from today's Gyeongju and unified the peninsula in the seventh century. Even if Buyeo languages affected the formation of Korean language, they played only a secondary role. Although it is unknown when Han Koreans became dominant in the southern part, probably they came from Manchuria with bronze weapons and ruled indigenous rice planters.

In the fourth century, Goguryeo invaded the peninsula and destroyed Chinese colonies in 314. Goguryeo moved its capital from Ji'an in Jilin province to Pyongyang in 427 opening the three kingdoms period. Beakje appeared from Mahan and dominated today's Gyeonggi and Chungcheong, and later expanded to Jeolla province. Silla appeared from Jinhan in eastern Gyeongsang province. The former Byeonhan area in southern Gyeongsang was not unified but sustained a loose association called Gaya confederacy. This area was annexed to Silla in the sixth century. After struggling of three kingdoms that lasted for four centuries, Silla finally defeated Goguryeo and Baekje under the support from the Tang dynasty and unified the Korean peninsula in 668. Since then, Silla's language and culture became dominant to form the ethnic identity of Koreans. It is supposed that cultures of Goguryeo, Beakje, and Gaya mostly disappeared during the unified Silla period (668–935). There were a large number of Chinese people in Korea during the unification war but they either returned to China or assimilated in unified Silla. Thus, it can be said that Koreans as an ethnic group was formed in this period.

2.8 Origin of Mongolic Peoples

The Mongolic peoples refer to ethnic groups that speak Mongolic languages. The majority in Mongolia today is the Khalkha Mongols. Mongolic peoples are assumed to be descendants of Genghis Khan who founded the Mongol empire in the thirteenth century. Because own historical records by Mongols started at that time, we need to rely on Chinese literatures to know about Mongols before Genghis Khan. Ancient ethnic groups that were likely to be the proto-Mongolic speakers include Khitan that established the Liao dynasty (916–1125), Xianbei that founded the Sui (581–618) and the Tang (618–907), and Xiongnu that fought against the Western Han (202 BC–9 AD). Although they were evidently nomadic peoples, Robbeets et al. (2021) pointed out that the proto-Mongols derived from farmers cultivating millets in the West Liao River region about 9000 BP. According to their linguistic analysis, the proto-Altaic was formed about 6800 BP, and the proto-Turkic separated from the proto-Mongolo-Tungusic about 4500 BP. These Altaic groups were farmers throughout the Neolithic Period, and started the nomadic lifestyle in the Bronze Age that began 4000 BP.

2.9 Summary and Conclusion

This chapter discussed the origin of the Chinese, the Japanese, Koreans, and the Mongols. While the Chinese is the offspring of the Dene-Caucasian group that was the first Homo sapiens group inhabited tempered Eurasia, the Japanese, Koreans, and the Mongols derived from the Eurasiatic superfamily that settled in the western shores of the northern Pacific during the paleolithic age. Although the homeland of the Sino-Tibetan language family is controversial, it is likely that the leaders of the Yellow River civilization were Sino-Tibetan speakers. The Chinese writing system goes back to the oracle bone scripts in the Shang dynasty, which implies that China is the only civilization that sustains the continuity from the cradle. Although the Han Chinese is assumed as one ethnic group, the difference between dialects is larger than that between European languages. This is why Taiwan is seen as a multi-ethnic society. Taiwan was a frontier occupied by a sparse indigenous population until the sixteenth century. The Hoklo people led by Koxinga came in the seventeenth century, followed by the Hakka in the nineteenth century and the Mainlanders in the late 1940s.

It was assumed in this chapter that the wet rice planting was transmitted to Korea and Japan by the Sino-Tibetan speakers because it is unlikely that the proto-Koreans or proto-Japanese lived on rice in Shandong peninsula in the early first millennium BC. This means that the southern part of Korean peninsula and western Japan were Chinese colonies during the Neolithic age, and that Koreans and the Japanese were late comers. It was hypothesized that the proto-Japanese holding metal weapons conquered western Japan in the second century BC. It is unknown when Han Koreans and various Altaic groups moved into Korea. The direct ancestor of Koreans was the Jinhan that settled in Gyeongsang province. While the Jinhan people founded Silla and unified Korea in the seventh century, languages and cultures of other ethnic groups disappeared.

According to Robbeets and her coauthors, the Transeurasian family including Turkic, Mongolic, Tungusic, Korean and Japanese languages emerged in the West Liao River region about 9000 BP. Greenberg discussed that the family was divided into the Altaic group and the Japanese-Korean-Ainu group. The Altaic group transitioned to the nomadic lifestyle in the Bronze Age that began 4000 BP. Those equestrian people often founded conquering dynasties such as Liao, Shui, Tang, Yuan by the Mongolic peoples or Jin and Qing by the Tungusic peoples. The Turkic peoples advanced to the west and threatened Europe by founding the Timurid Empire or the Ottoman Empire, although these are out of the range of this book.

References

Sagart, Laurent, Roger Blench, and Alicia Sanchez-Mazas. 2005. Introduction. In *The Peopling of East Asia: Putting Together Archaeology, Linguistics and Genetics*, ed. Laurent Sagart, Roger Blench, and Alicia Sanchez-Mazas. Routledge Curzon.

Diamond, Jared. 1998. In search of Japanese roots. *Discover Magazine*, June 1, 1998.

Robbeets, Martine, et al. 2021. Triangulation supports agricultural spread of the Transeurasian languages, *Nature*, November 10, 2021.

Kudo, Susumu. 2005. *Nihongo-ha Doko-kara Umareta-ka – Nihongo Indo-yoroppago Doitsu Kigensetsu.* Best Sinsho (in Japanese).

Tanaka, Katsuhiko. 2021. *Kotoba-ha Kokkyo-wo Koeru – Nihongo, Ural-Altai-Tsuran Shugi.* Chikuma Sinsho (in Japanese).

Shinoda, Ken'ichi. 2020. *Shinpan: Nihonjin-ni Natta Sosentachi – DNA-ga Kaimeisuru Tagenteki Kozo.* NHK Books (in Japanese).

Manning, Patrick. 2006. Homo sapiens populates the earth: A provisional synthesis, privileging linguistic evidence. *Journal of World History* 17 (2): 115–158.

Bellwood, Peter. 2005. Examining the farming/language dispersal hypothesis in the East Asian context. In *The Peopling of East Asia: Putting Together Archaeology, Linguistics and Genetics*, ed. Laurent Sagart, Roger Blench, and Alicia Sanchez-Mazas. Routledge Curzon.
Kazama, Kiyozo. 1993. *In'ougo-no Kokyo-wo Saguru*. Iwanami Shoten (in Japanese).
van Driem, George, 2005. Tibeto-Burman vs Indo-Chinse: Implications for population geneticists, archaeologists and prehistorians. In *The Peopling of East Asia: Putting Together Archaeology, Linguistics and Genetics*, ed. Laurent Sagart, Roger Blench, and Alicia Sanchez-Mazas. Routledge Curzon.
Starosta, Stanley. 2005. Proto-East Asian and the origin and dispersal of the languages of East and east Asia and the Pacific. In *The Peopling of East Asia: Putting Together Archaeology, Linguistics and Genetics*, ed. Laurent Sagart, Roger Blench, and Alicia Sanchez-Mazas. Routledge Curzon.
Diamond, Jared. 1997. *Guns, Germs, and Steel: The Fates of Human Societies*. W. W. Norton.
Okuno, Masao. 2000. *Tetsu-no Kodaishi 3 – Kibabunka*. Hakusuisha (in Japanese).
Hayashi, Toshio. 2017. *Kobo-no Sekaishi: Sukitai-to Kyodo – Yuboku-no Bunmei*. Kodansha (in Japanese).
Ho, Ping-Ti. 1959. *Studies on the Population of China, 1368–1953*. Harvard University Press.
Greenberg, Joseph H. 2000. *Indo-European and Its Closest Relatives, the Eurasiatic Language Family, Volume 1. Grammar*. Stanford University Press.
Lee, Hong Kyu. 2020. Origin of Korean people and DNA tracking. *Hanguk Siberia Yeongu* 24 (2): 113–160.
Egami, Namio. 1967. *Kibaminzoku Kokka – Nihon Kodaishi-heno Approach*. Chuko Sinsho (in Japanese).
Osada, Natsuki. 2010. *Shinko: Yamataikoku-no Gengo – Yayoigo Fukugen*. Gakuseisha (in Japanese).
Fujio, Shin'ichiro. 2015. *Yayoi Jidai-no Rekishi*. Kodansha (in Japanese).
Harari, Yuval Noah. 2011. *Sapiens: A Brief History of Humankind*. Harvill Secker.
Saito, Naruya. 2017. *Kaku DNA-de Tadoru Nihonjin-no Genryu*. Kawade Shobo Sinsha (in Japanese).
Kito, Hiroshi. 2007. *Zusetsu: Jinko-de Miru Nihonshi – Jomon Jidai–kara Kinmirai Shakai-made*. PHP Interface (in Japanese).
Sato, Yoichiro. 2002. *Ine-no Nihonshi*. Kadokawa Sophia Bunko (in Japanese).
Kimura, Shigemitsu. 2010. Genshi. In *Nihon Nogyoshi*, ed. Kimura Shigemitsu, 11–38. Yoshikawa Kobunkan (in Japanese).
Pang, Min Kyu. 2018. Saengmul inryuhak jaryo-ro bon hangugin giweonmunje-e gwanhan yeongu. *Inmun Sahoe 21* 9 (3): 1211–1226 (in Korean).
Choi, Ki-Ho. 2004. Hangugeo gyetong-gwa hyeongseong-eui munjejeom. *Mongolhak* 16: 241–267 (in Korean).

Chapter 3
Socioeconomic and Demographic Transitions

The most important transitions in human history were the shift of production modes from hunting and gathering to farming, and from farming to manufacturing. Thus, the most basic historical periodization distinguishes a hunter-gatherer society, an agrarian society, and an industrial society. The neolithic revolution and the industrial revolution enabled dramatic increases in sustainable population and accompanied thorough changes in political, economic, and cultural systems. The demographic transition theory emphasizes changes in the levels of fertility and mortality. It is presumed that fertility increased with the transition to an agrarian society because the child labor force was more important than in a hunter-gatherer society. It is also supposed that mortality increased due to the risk of infectious diseases in densely inhabited villages (Caldwell 2006; Harari 2011).

While the planting of millet, barley, and bean began 9500 BP in the Yellow River basin, the wet rice planting in the Changjiang basin started at about the same time or even earlier. China was one of four cradles of civilization and held high technological levels in most fields including metalware, pottery, textile, architecture, and civil engineering. Most important inventions in the human history such as paper, gunpowder, printing, and the compass were achieved in China. Apparently, China was the wealthiest country in Eastern and Southeastern Asia until the eighteenth century.

The wealth of China stimulated invasions of surrounding nomad peoples. Sometimes those horse riders built a dynasty and ruled the Han Chinese. However, the cultural level of the Han was so high that they were never conquered culturally or gave up their legacy. Instead, some conquerors were culturally occupied by the Han Chinese and lost their own language and ethnic identity. Although imperial families of Sui (581–618) and Tang (618–907) originated from the Xianbei, an Altaic group, they were already sinocized at the foundation of dynasty and had no intention to preserve own language or other cultural legacies. The Liao dynasty (946–1125) built by the Khitan did not occupy the central China and was relatively free from sinocization. Rulers of the Jin dynasty (1115–1234) of Manchurian origin and the Yuan

T. Suzuki, *Comparative Population History of Eastern Asia*,
https://doi.org/10.1007/978-981-99-9367-3_3

dynasty (1271–1368) of Mongolian origin experienced some sinocization during they ruled China for a century but returned to their homelands before they lose the ethnic identity. The Manchu founders of Qing dynasty (1644–1912) forced the queue hairstyle, defined the Manchu language as official, and limited the number of the Han Chinese in the top-ranked bureaucrats. Despite those ethnocentric policies, sinocization of the Manchu elites proceeded continuously for centuries. Although there are ten million people self-proclaiming the Manchu ethnicity today, nobody can fluently speak the Manchu language. In this way, the Han Chinese always won in cultural competitions and absorbed surrounding ethnic groups.

Although primitive agriculture was conducted in the Jeulmun pottery period (10,000–3500 BP) in Korea or the Jomon period in Japan, it is controversial if those were agrarian societies. This book views that the use of wet rice paddy marks the beginning of an agrarian society. Thus, Korea and Japan transitioned to an agrarian society 3000 BP and 2900 BP, respectively. The irrigated rice cropping requires a large-scale transformation of the environment and a revolutionary change of thoughts. The formation of social strata and the beginning of warfare were also accompanied by the transition to an agrarian society.

The Korean peninsula is in the middle of military powers of China, Mongolia, Manchuria, Russia, and Japan. Due to this geopolitical position, Korea suffered from military invasions many times in its history. While the Joseon dynasty (1392–1910) heavily depended on China, Koreans sustained their ethnic identity and developed Small-sinocentrism, the Korean version of Sinocentrism. To the contrary, Japan was isolated from the continent and was never conquered by a foreign power until 1945. Mongolian expeditions in the fourteenth century failed by the efforts of Japanese warriors and the oncoming of typhoons.

The industrial revolution allowed a spectacular growth of production and mortality declined rapidly thanks to improvements in nutrition and public health institutions. Medical inventions such as sulfonamides and antibiotics facilitated the global population growth after World War II (WWII). The value of child labor declined as a family lost its function as a production unit. The development of social welfare institution supporting the elderly people also promoted fertility decline. In the post-industrial stage, fertility declined to the below replacement level. As mentioned above, low fertility issue is especially serious in Eastern Asia.

It is assumed in this book that the transition to an industrial society is equivalent to the beginning of the modern era in Eastern Asia. Unlike ern Europe that modernized spontaneously, non-European regions were forced into modernization by the western impact. Most history textbooks in China define that the modern era began with the First Opium War in 1840–42. It is common in Japanese history to define that the modern period began with the Meiji Restoration in 1868 and ended with Japan's defeat in the Pacific War in 1945. The range of modern Korean history is also well defined that it began with the Ganghwa Island Incident in 1875. Recent history textbooks in Taiwan defines the period of Japanese rule (1895–1945) as the modern period.

When the contemporary period is distinguished from the modern period, it is common in Eastern Asia that the conclusion of WWII is chosen as a marker. Thus, the surrender of Japan in 1945 marks the beginning of the contemporary period in Japan, Korea, and Taiwan. In China, the foundation of the People's Republic of China in 1949 marks the beginning of the contemporary period.

3.1 Historical Periods

Table 3.1 summarizes the division of historical periods in Eastern Asia since 2000 BC. Jomon, Yayoi, and Kofun periods in Japan are based on archaeological studies and written records are very scarce. As mentioned above, Jomon and Yayoi are names of pottery styles. While the oldest Jomon pottery goes back to 16,000 BP, the beginning of Yayoi period is marked by the wet rice paddy in northern Kyushu in 900 BC. The expansion of Yayoi culture was relatively moderate, and the Jomon and Yayoi lifestyles coexisted for a long time in Japan. Kofun means an "ancient burial mound" and a specific style called "zenpo-koen-fun," a combination of a trapezoid and a circle, spread in whole Japan other than Hokkaido and Okinawa between the third and seventh centuries. It is supposed that the prevalence of zenpo-koen-fun exhibits the process of unification by the Yamato people that based on the Nara basin. There were various ethnic groups in the Japanese archipelago, as in the Korean peninsula, that were absorbed or extinguished during the unification. Among rivals of the Yamato, the Emishi people occupied northeastern Japan are thought to be the offspring of the Jomon people and spoke proto-Ainu languages (Diamond 1998). The Hayato people in southern Kyushu also might have been non-Japanese speakers but the detail is unknown. There must have been other ethnic groups that disappeared without being recorded in the history.

The *Nihon Shoki*, one of the oldest history books in Japan, was compiled in 720. Although the book contains the history for more than a thousand years, contents before the sixth century are unreliable. Thus, the Jomon, the Yayoi, and most of the Kofun period should be seen as prehistory. However, there remains a detailed historical record of Japan in the early third century written by the Chinese. The "Biography of Dongyi" in the *Book of Wei* includes a record of Japan in addition to various ethnic groups in Korea and Manchuria. According to the book, there was a civil war in Wa (Japan) in the second century, which ended when a shaman named Himiko was selected to be the queen. In 239, Himiko sent envoys to the Wei dynasty in northern China and was given favorable treatment. Although a king was enthroned after Himiko died in 247 or 248, people did not obey, and a war broke out again. A girl named Toyo, a relative of Himiko, was selected as the queen and the peace was restored. This record includes the geographical information of Japan seen from the Chinese. However, the description is too brief and inaccurate to determine the location of Himiko's palace. The location of Yamatai, Himiko's domain, remains as the most controversial mystery in the history of Japan (Kobayashi 2022).

Historical periods between the sixth and the nineteenth centuries are named after the location of government. While emperors and empresses resided in Asuka at the southern end of Nara basin, envoys were exchanged between Japan and the Sui

Table 3.1 Historical periods in Eastern Asia

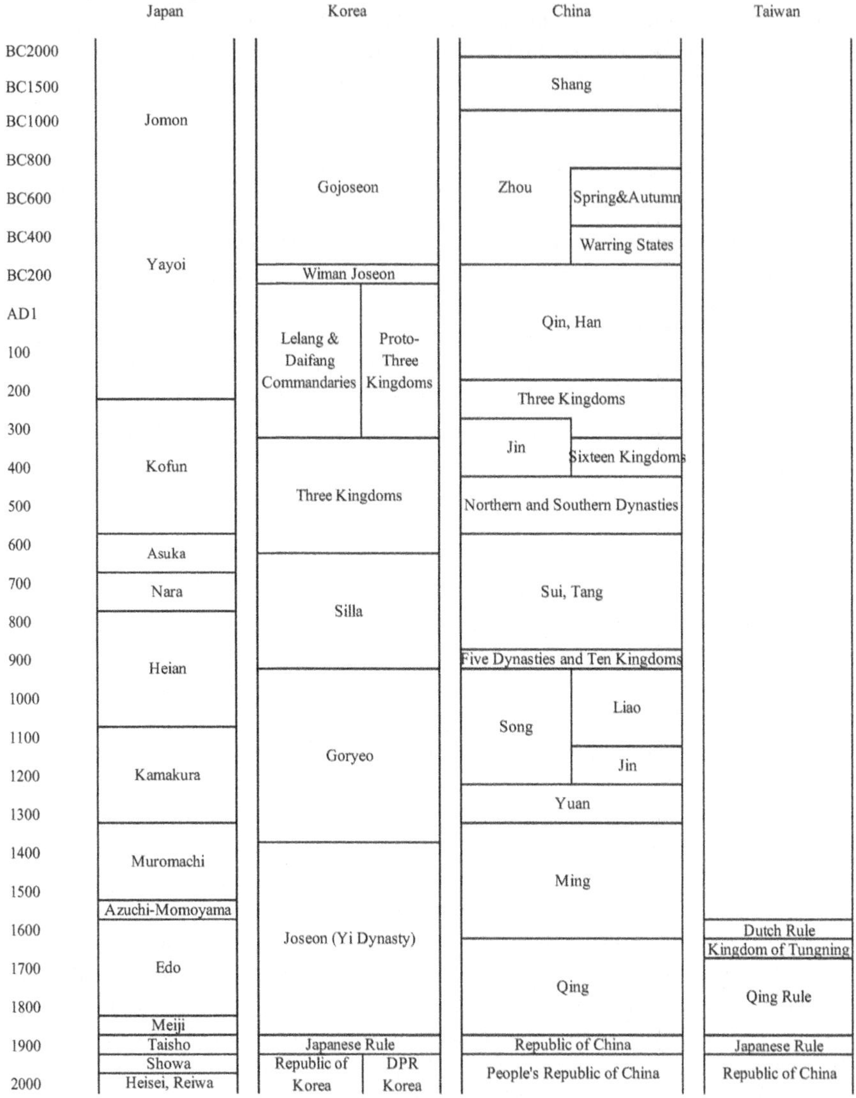

dynasty (607–608), and an expedition force was sent to help Baekje (663). In 710, the capital was moved to Heijo-kyo, today's Nara city. During the Nara period (710–794), the Chinese style bureaucracy with an emperor at the top was developed, and the enlargement of cultivation was promoted. In the Heian period (794–1192) when the capital was in Heian-kyo, today's Kyoto city, the power of emperor declined, and the political system shifted to the aristocracy of the Fujiwara clan. During this period, the cultural influence of China and Korea decreased, and the original Japanese culture developed in various fields such as architectures, poems, novels, paintings, and clothes.

One of the most important historical changes in Japan occurred when Minamoto no Yoritomo (1147–99) founded the first "bakufu" (shogunate) in Kamakura in 1192. This incident implies the transition from aristocracy to feudalism, from a government of aristocrats to that of samurai warriors, and from ancient to medieval time. Feudalism is a loose integration of feudal lords that had their own military powers and local laws. In premodern Europe and Japan, the warrior classes of knights and samurais, respectively, took political power and governed the nation. Feudalism emerged under the declining authority of the central empire and eventually evolved into absolutism both in Europe and Japan (Eisenstadt 1996). By contrast, China and Korea had agrarian bureaucracies, in which bureaucratic aristocrats such as scholar-officials and yangban governed the nation in a more centralized and top-down way than feudal societies did. Civilian control was established in China and Korea, and warriors did not hold political power (Cumings 2005).

It is often discussed that feudalism was adaptive to later historical developments of modernization and industrialization. Umezao (1967) asserted that Japan and Europe were special places where "autogenic succession" proceeded to prepare the bourgeoisie and capitalism. Wittfogel (1959) pointed out the similarity of the decentralized political system and the protection of private property, and suggested these characteristics promoted capitalism. Goode (1963) compared Japanese feudal society with Chinese familial society. He listed Japanese family characteristics that contributed to the industrialization process such as differential family norms and behavior by class and region, the position of the family as the last chain of political hierarchy, no obligation for successful men to support a wide range of blood relatives, and continuous capital cumulating by merchants rather than desiring to be a mandarin. According to Hayami (2015), the establishment of decentralized society was the precondition of the shift to "economic society" in which the economic principles became independent from other social phenomena. In China and Korea, however, highly centralized societies perpetuated and resulted in "fossilization," and eventually colonized by imperial powers.

The Kamakura shogunate could defend against the Mongolian invasion in 1274 and 1281 but could not give rewards to vassals that fought with their own costs. The disaffection against the shogun rose among impoverished samurais and rebellions broke out frequently. The turmoil was further heightened when the shogunate interfered in the problem of imperial succession, and the Kamakura government finally collapsed in 1333. Ashikaga Takauji (1305–58) closed the civil war and built his shogunate in Kyoto city in 1388. Since the mansion of shogun was at Muromachi town, the period until the last shogun was banished from Kyoto is called the Muromachi period (1336–1573). After the time of peace and wealth for a century, the Onin war (1467–77) opened the warring state period for more than a century. The authority of shogun fell and many feudal loads called "daimyos" fought each other. The long civil war was finally ended by Oda Nobunaga (1534–82) and Toyotomi Hideyoshi (1537–98). Because Nobunaga based at the Azuchi castle in today's Shiga prefecture and Hideyoshi's Fushimi castle was at Momoyama town in Kyoto city, this period is called the Azuchi-Momoyama period (1573–1603).

After the death of Hideyoshi, Tokugawa Ieyasu (1543–1616) opened his shogunate in Edo, today's Tokyo, and the Edo period (1603–1868) started. It is usual in Japan to distinguish the early modern period from the medieval period. In that case, the Edo period is entirely included in the early modern period. A regime similar to the absolute monarchism in Western Europe was established by strengthening the power of shogun while weakening that of feudal lords. The Edo period also saw a transition to a "peasant society." Agricultural production in ancient and medieval eras was executed by large scale units including lateral kinship members or economically dependent workers such as slaves or serfs. In the early modern era, however, Western Europe and Eastern Asia transitioned to peasant societies in which most agricultural production was executed by small family units. The transition took place in the seventeenth century in Japan when the expansion of farmland reached its limit and an improvement in land productivity was required.

The Meiji restoration (1868) ended the rule of samurai, and a centralized state with an emperor at the top was built. Meiji (1868–1912), Taisho (1912–26), Showa (1926–89), Heisei (1989–2019), and Reiwa (2019–) are regnal era names assigned for each emperor. Under the western impact in the late nineteenth century, Japan opened the country and promptly achieved the modernization and industrialization. In the world of colonialism and imperialism, Japan won in the First Sino-Japan War (1894–95) and the Russo-Japanese War (1904–05), occupied a German concession in Qingdao (1914), but finally defeated by the US in WWII. After the occupation by the Allies (1945–52), Japan was built into the capitalist camp under the cold war situation, and collaborated with the US, the Republic of Korea, and the Republic of China (Taiwan). Japan experienced the rapid economic growth in the 1960s and 1970s to be the second largest economy in the world.

While the English word "Korea" came from the Goryeo dynasty (918–1392), the Republic of Korea ("Daehan Minguk" in Korean) has "Han" in the nation's name, and the Democratic People's Republic of Korea ("Joseon Minjujueui Inmin Gonghwaguk") self-proclaims "Joseon." The Yi dynasty (1392–1910) also used "Joseon" as the country name. "Gojoseon," literally "ancient Josen," in Table 3.1 is a term to distinguish from later Joseons, and consists of Dangun Joseon, Gija Joseon, and Wiman Joseon. Dangun appears as the legendary founder of Joseon in *Samguk Yusa* (Memorabilia of the Three Kingdoms) written in the thirteenth century. Shima Qian wrote in *Shiji* that Gija ("Jizi" in Chinese), a sage in the Shang dynasty, went to Korea and founded a country. However, it is difficult to accept them as real history, and the period before Wiman Joseon should be seen as prehistoric period. As mentioned above, the Han dynasty wiped out Wiman Josen in 108 BC and founded four Chinese colonies including Lelang commandery that lasted until being destroyed by Goguryeo in 314. According to the "Biography of Dongyi" in the *Book of Wei*, in the early third century, there were Buyeo and Goguryeo in Manchuria, Chinese colonies in northwestern Korea, Okjeo and Ye-Maek in northeastern Korea, and Han Koreans in southern Korea. This period is called the "proto-three kingdoms period" (108 BC–314 AD), which was the prelude to the "three kingdoms period" (314–668) when

Goguryeo, Baekje, and Silla fought each other. The coalition force of Silla and Tang defeated Baekje in 660 and Goguryeo in 668. As already discussed, the ethnicity of Koreans was formed during the unified Silla period (676–935).

When the Silla declined, Wang Geon (877–943) excluded his rivals and built the Goryeo dynasty (918–1271). Because the Goryeo defined the Buddhism as the state religion, priests and monks had political influences. While various cultural achievements such as the compilation of Buddhist texts, the creation of Goryeo ware, the introduction of the imperial examination of China were made in the Goryeo period, the country suffered from invasions of the Kittan troops, the Japanese pirates, and especially the Mongolian army. Ögedei Khan, the second emperor of the Mongol empire, started the war against the Jin dynasty (1115–1234) in 1230, and invaded to Korea in 1231. While the king of Goryeo fled to Ganghwa island, the Mongolian force ravaged the whole peninsula. The Goryeo finally surrendered in 1259, and Mongolian officials were dispatched to supervise the Goryeo people. An enormous number of Korean workers and soldiers were mobilized for expeditions to Japan in 1274 and 1281.

When the Yuan (1271–1368) was replaced by the Ming (1368–1644) in China, a Korean general Yi Seong-gye (1335–1408) overthrew the Goryeo and built the Joseon dynasty (1392–1910). The kingdom lasted for more than 500 years, longer than any dynasty in China or shogunate in Japan. The Joseon dynasty excluded the Buddhism and determined Confucianism as the state religion. As a result, bureaucrats had the political influence, and efforts were made to sinocize the political system. Japan's invasion in 1592–98 was repelled owing to operations of the Korean navy and reinforcements from China. When the Qing (1644–1912) replaced the Ming, Korea was forced a subordinate position. Korean intellectuals considered that Manchu rulers of the Qing were barbarians, and Small-sinocentrism asserting that Korea is the legitimate successor of the Chinese civilization developed. There is a disagreement on the transition to the peasant society in Korea. While Miyajima (1994) discussed that the transition completed in the seventeenth century in Korea, Nakamura (2004) concluded that it completed in the late eighteenth century. According to Shikata (1976), the percentage of slaves called "nobi" in Daegu area decreased in the eighteenth century but rose again in the nineteenth century. When the official abolishment of slavery was attempted in the Gabo reform (1894–96), the yangban class disapproved it (Lee 2020). This implies that there were a considerable number of slaves in the nineteenth century and that the transition to a peasant society was less complete in Korea than in Japan.

Japan occupied Ganghwa island in 1875 and forced Korea to open the country. According to the treaty concluded in the next year, Korea opened Busan and other two ports, which marked the beginning of the modern period of Korea. After the Russo-Japanese War (1904–05), Korea was eventually annexed to Japan. During the Japanese rule (1910–45), the agricultural development in the south and industrialization in the north were promoted. When Japan surrendered to the Allies in 1945, Korea was divided at the 38th degree line, and the US occupied the southern part while the Soviet Union dominated the northern part. The division resulted in the foundation of two nations, the Republic of Korea and the DPR Korea, in 1948. The

Korean War (1950–53) broke out with the invasion of the North, but the United Nations Command landed at Incheon repelled the North Korean force. When the UN coalition force approached to the border of China, the People's Liberation Army participated and pushed back the frontline to the 38th degree line. Leaders of China, the USSR, and the US did not want to continue the war and concluded the Armistice Agreement in 1953.

Although the South Korean economy stagnated during the 1950s, successful developmental policy under the presidency of Park Chung-Hee (in office 1963–79) brought about the "miracle on the Han River." While generating serious human rights problems, Park's developmental dictatorship enabled Korea to transition swiftly from an agrarian to a highly industrialized society. Both the US and Japan aided Korea eagerly under the cold war situation. While President Chun Doo-Hwan (in office 1980–88) succeeded the dictatorship, President Roh Tae-Woo (in office 1988–93) had to promise democratic reforms during the presidential election in 1987. The high economic growth rate of Korea was disrupted by the Asian financial crisis in 1997. Although the Korean economy recovered quickly from the damage, growth has been slower than before the crisis.

The economic growth of the north was faster than the south in the 1950s and 1960s owing to the infrastructure built during the Japanese era. However, most political and economic leaders cooperated to the Japanese rulers were in the south. The case of divided Korea suggests that capitalism with good human resources is better than socialism with good material resources for economic development in a long-term. It is estimated that the Republic of Korea overtook the DPR Korea in the mid-1980s in terms of income per capita (Cumings 2005). After the collapse of the former Soviet Union, DPR Korea suffered economic difficulty. The flood in 1995 seriously damaged agricultural production, and a large number of people died from the long-term famine in 1996–2000.

Although Sima Qian wrote about the Xia dynasty, the predecessor of the Shang dynasty, the historical reality of the Xia is still controversial. On the other hand, the reality of the Shang, also known as the Yin dynasty, is well-established by oracle bone scripts discovered from ruins in Henan province. In 1046 BC, the Shang lost in the war against the Zhou dynasty (1046–26 BC) that based on Shaanxi province. Although the political regime of the Zhou was feudalism, Umezao (1967) asserted that it was different from feudalism of Western Europe and Japan that formed bourgeois class and prepared capitalism. In a rebellion in 771 BC, King You was killed, and the capital was moved to today's Luoyang city in Henan province. The power of the Zhou declined and the Spring and Autumn period (770–403 BC), in which independent lords fought each other, began. After struggles for centuries, seven warring states remained. The era until the unification by the Qin is called the Warring States period (403–221 BC). Before the expansion of Sinitic language group, various ethnic groups speaking Tibet-Burman, Hmong-Mien, Tai-Kadai, Austro-Asiatic, and Austronesian languages distributed in southern China. Sinocization of the south already started before the unification by the Qin. Starosta (2005) inferred that the Chu kingdom in the lower Changjiang was a country in which a Han governing class ruled a primarily Hmong-Mien population.

The Qin dynasty defeated six warring states and unified whole China in 221 BC. The Qin fell due to a peasant rebellion in 206 BC and replaced by the Han (206 BC–220 AD). The toughest rival of the Han was the Xiongnu, nomadic Mongolians inhabited the Mongolian plateau. Although the Han was forced an inferior position to the Xiongnu at the beginning, the seventh emperor Wu (reign 141–87 BC) took an expansionist policy and his expedition forces defeated the Xiongnu several times. As a result, the Han had superior influence in Xinjiang and central Asia. The Han also destroyed Wiman Joseon and founded four commanderies in northwestern Korea.

After the ruin of Han, China was shared by the Cao Wei (220–265), the Shu Han (221–263), and the Eastern Wu (222–280). This three kingdoms period (220–280) ended when the Western Jin (265–316) unified China in 280. In the fourth century, barbarians in the north and the west became more and more active and often invaded Middlefield. After the Xiongnu defeated the Western Jin in 316, northern China entered another warring state period called "Sixteen Kingdoms of the Five Barbarians" (304–439). Five barbarians are Xiongnu and Xianbei of Mongolian origin, Qiang and Di of Tibetan origin, and Jie supposed to have been speakers of an Indo-European language. Sixteen kingdoms built by these ethnic groups fought each other for the hegemony of northern China. In southern China, the royal family of the Western Jin fled to today's Nanjing city and built the Eastern Jin (317–420). In 420, the Eastern Jin was replaced by the Liu Song (420–479). The Northern Wei (386–534) of Xianbei origin finally unified northern China in 439 and opened the period of "Northern and Southern Dynasties" (439–589).

In 589, the Sui (581–618) in the north ruined the Southern Chen (557–589) and unified China. The Sui dynasty spent too much power for building the Grand Canal and expedition to Goguryeo and was replaced by the Tang (618–907). The Tang force stayed in the Korean peninsula after defeating the Baekje and Goguryeo but retreated in 678. The Tang lost in the Battle of Talas (751) against the Islamic forces and withdrew from central Asia. After the Tang collapsed by a peasant rebellion, China fell into a period of turmoil called "Five Dynasties and Ten Kingdoms" (907–960).

China experienced major socioeconomic changes during the Song period (960–1279). First of all, the establishment of imperial examination system did not permit aristocrats to survive as a social class. The succession of social status became impossible, and meritocracy was enforced as the principle of social mobility. The social stratum was simplified and consisted of scholar-gentlemen that could produce bureaucrats and ordinary people that could not (Okamoto 2013a). Economically, the Changjiang delta was developed, multi-cropping became possible, and the food production grew dramatically (Ho 1959). Coal replaced wood as fuel, and mass-production of consumer goods such as metal wares, porcelain, and tea started. Money economy prevailed widely, and banknote was used among traders. These changes are summarized as the "first commercial revolution" in China (Tetsuyama 2000; Okamoto 2013b).

The Song dynasty had to face very powerful enemies in the north and the west. The Khitan of Turkic origin attacked the Song in 1004 and forced a subordinate position, although they did not demand a large area of Song's domain. In 1044, the Song had to pay a huge amount of silk and silver due to the pressure of Western Xia (1038–1227).

The Song collaborated with the Jin dynasty (1115–1234) in Manchuria and defeated the Khitan state Liao in 1121. However, the Song had to retreat to southern China due to the invasion of the Jin in 1127. The imperial court of Song was originally in today's Kaifeng in Henan province but moved to Hangzhou in Zhejiang province at the invasion. The Song until 1127 is referred to as the "Northern Song," while the "Southern Song" is used for the period after the retreat. The Mongol empire defeated the Jin in 1234 and occupied northern China. The Yuan dynasty (1271–1368) of the Mongol defeated the Southern Song in 1279 and dominated whole China.

While the Mongol empire held the largest contiguous land in the human history, its rule of whole China lasted only for nineteen years. Since the Mongol was not good at battles on the water, Khubilai's expedition to Japan and Indonesia eventually failed. The conflicts over throne succession in the fourteenth century seriously undermined the national power of the dynasty. After a series of nationwide rebellions, Zhu Yuanzhang (1328–98) excluded rivals and built the Ming dynasty (1368–1644).

There were several agricultural inventions during the Ming period such as the increase of draft animals, the construction of irrigation systems, and the spread of fertilizer. However, the most important change was the introduction of American food plants. Maize, potatoes, sweet potatoes, and peanuts were planted since the early sixteenth century in the southwestern and northeastern parts of China and stimulated a large flow of settlers and expansion of agricultural fields (Ho 1959). On the other hand, the Changjiang downstream transitioned from the rice monoculture to the production area of commodity crops such as cotton, silk, and tea. These products in addition to porcelain became popular in abroad and were traded in a large scale. As a result, a huge amount of silver flew into China from the new continent and Japan. Those changes are referred to as the "second commercial revolution" in China (Okamoto 2013b).

When Japan attacked Korea in the end of the sixteenth century, the national power of the Ming dynasty was strong enough to push back the coalition force of Japanese daimyos. In the seventeenth century, however, the Ming lost its power and rebellions broke out nationwide. When a rebel army led by Li Zicheng (1606–45) entered Beijing in 1644, the last emperor Chongzhen (reign 1627–44) committed suicide and the reign of the Ming dynasty was terminated. Shortly after that, Manchu forces expelled Li Zicheng from Beijing and the rule of the Qing dynasty of China (1644–1912) began.

Owing to the maritime bans enforced to contain the Ming loyalists in Taiwan, the flow of silver stopped, and the economy stagnated. After the annexation of Taiwan in 1683, international trade recovered, and the economy started growing fast. Trade with European countries went well during the eighteenth century, and China experienced the dramatic simultaneous growth of population and economy. In the nineteenth century, the growing demand of funds due to the industrial revolution in the UK made it difficult to export silver. The British began to export opium from Indian colonies to China to compensate the trade deficits caused by imports of Chinese tea. This was the background of the Opium War.

In the modern era of China, the Qing lost in a series of incidents including the First Opium War (1840–42), the Arrow War (1856–60), the First Sino-Japanese War

(1894–95), and the Boxer Rebellion (1900–01). The Qing dynasty finally fell in 1912, and the Republic of China was founded. However, the central government could not control the nation, and many warlords took power in each local area after the death of Yuan Shikai (1859–1916). The conflict between nationalists and communists was intensified in the 1920s. Japan conquered Manchuria and founded Manchukuo, a puppet state, in 1932. The Second Sino-Japanese War broke out in 1937 lasted until the surrender of Japan in 1945.

The Chinese civil war after the retreat of Japan ended in the victory of the communists, and the Kuomintang (KMT) escaped to Taiwan in 1949. In the People's Republic of China, the economy began to recover from the damage of wars, and the first five-year plan (1953–57) was achieved successfully. However, Mao Zedong's ambitious plan of the Great Leap Forward (1955–62) ended in a disastrous failure, causing a catastrophic collapse in agricultural and industrial productions. Millions of people died from hunger during the great famine in 1959–61. When Mao's reputation declined, he started the Cultural Revolution (1966–76). Political conflict and anti-capitalistic policy resulted in a serious economic recession. Economic reforms led by Deng Xiaoping (1904–97) opened the door to economic growth after the 1980s. Chinese overseas played an important role in introducing capital and technology in the early stage. While the GDP of China overtook that of Japan in 2011 to become the second largest economy in the world, per capita income is still far below that of Japan, Korea, and Taiwan.

Table 3.2 summarizes the timing of various transitions in Eastern Asia. As mentioned above, there is a possibility that the agriculture started earlier in the Changjiang basin than in the Yellow River basin. The arrival of the wet rice cropping is taken as the indicator of agrarian society in other three societies. The oldest oracle bone script goes back to the fourteenth century BC in the Shang dynasty. The Chinese did not leave a detailed description on the society of indigenous Taiwanese before the Dutch period. Historical events in Korea were continuously recorded by the Chinese since the era of Wiman Joseon (194–108 BC). The oldest historical book written by Koreans that remains today is *Samguk Sagi* (*History of Three Kingdoms*) completed in 1145. As discussed above, contents before the sixth century in the *Nihon Shoki* is unreliable, but the *Book of Wei* has a detailed description about Japan in the third century.

Miyajima (1994) wrote that the transition to a peasant society in China completed in the early Ming period. Nakamura (2004) discussed the transition in Taiwan took place in the nineteenth century. While Miyajima and Nakamura agree with the timing of transition in Japan, they disagree about Korea. As already mentioned, incidents that mark the opening of modern era are the First Opium War (1840–42) for China, the annexation to Japan (1895) for Taiwan, the Ganghwa Island Incident (1895) for Korea, and the Meiji Restoration (1868) for Japan. Incidents that mark the contemporary era are the foundation of the People's Republic (1949) for China, and the surrender of Japan for other countries.

Table 3.2 Historical transitions in Eastern Asia

	China	Taiwan	Korea	Japan
Agrarian society	9500 BP, or earlier	Seventeenth century	3000 BP	2900 BP
Written record	Shang dynasty (fourteenth century BC)	Seventeenth century	Wiman Josen (second century BC)	Sixth century (third century)
Peasant society	Fifteenth century	Nineteenth century	Seventeenth to eighteenth century	Seventeenth century
Modern era	1840–1949	1895–1945	1875–1945	1868–1945
Contemporary era	1949–	1945–	1945–	1945–

3.2 Overview of Population Growth

It is estimated that the density of hunter-gatherer populations was at between 0.1 and 1 km^2. Since the population density of Northern/Western Europe was 40–60/km^2 in the mid-eighteenth century, the population grew by 100 times during several millennia of agrarian society (Livi-Bacci 2012). If we assume that the average length of agrarian society of the region was 3000 years, the average growth rate was 0.15%.

Table 3.3 shows the population density at the transition to an agrarian society and that to an industrial society in Japan, Korea, Taiwan, and China. The land area is that of today and may be different in the past. The beginning of agrarian society was defined by an introduction of wet rice cultivation. Such transitions took place around 7500 BC in China, 1000 BC in Korea, 900 BC in Japan (Bellwood 2005; Fujio 2015). In Taiwan, the irrigated paddy farming began in the sixteenth century when the Dutch East India Company invited Chinese peasants to cultivate the west coast. Then, the length of agrarian society was about 9300 years in China, 2900 years in Korea, 2800 years in Japan, and 280 years in Taiwan. It is reasonable to assume that Neolithic Eastern Asian societies had higher population densities than Europe thanks to warm and wet weather and rich maritime recourses. Thus, it is assumed here that the hunter-gatherer density was 1.0/km^2.

Historians agree that historical incidents that mark the beginning of the modern period are the First Opium War (1840–42) for China, the Meiji Restoration (1868) for Japan, the Ganghwa Island Incident (1875) for Korea, and the Annexation to Japan (1895) for Taiwan. Estimated populations close to these years were chosen in Table 3.3. For Korea, the high variant for the year 1850 estimated in Chap. 4 was used as the population at the beginning of modern period. Estimated population densities at the beginning of modern period suggest that rice cropping societies in Eastern Asia, especially Japan, achieved higher population densities than European countries. Although the density for whole China was relatively low, coastal provinces showed extremely high densities. For example, Jiangsu province with land area of

Table 3.3 Longterm population density and population growth

	Japan	Korea	Taiwan	China
Land area (km^2)	372,969	220,804	36,193	9,596,961
Transition to agrarian society				
Year	900 BC	1000 BC	1624	7500 BC
Population density	1.0	1.0	1.0	1.0
Transition to industrial society				
Year	1873	1875	1898	1845
Population (1,000)	33,301	13,869	2690	427,612
Population density	89.3	62.8	74.3	44.6
Maximum population				
Year	2010	2021	2019	2021
Population (1,000)	128,057	77,229	23,603	1,412,600
Population density	343.3	349.8	652.1	147.2
Annual growth rate (%)				
Agrarian society	0.16	0.14	1.58	0.04
Industrial society	0.99	1.18	1.81	0.68

Notes Population of Japan in 1873 was from Kito (2007)
Population of Korea in 1875 is the high variant in 1850 estimated in this book
Population of Taiwan in 1898 was from Mizoguchi (2008)
Population of China in 1845 was from Zhao and Xie (1988) cited in Ueda (1995)
Maximum population were from statistical bureaus of each country

106,740 km^2 had the population of 43 million in 1845 (Zhao and Xie 1988), which implies the density of 407.3/km^2.

Eastern Asian countries in Table 3.3 have already entered to the phase of long-term population decline. The population of Japan has been decreasing since 2010. According to the Statistics Korea, the combined population of the Republic of Korea and the DPR Korea hit the peak in 2021. The current population statistics by the Directorate General of Budget, Accounting and Statistics (DGBAS) showed that the maximum population in Taiwan was 23,603,121 in 2019. The National Bureau of Statistics reported that the population of China started declining in 2022. As shown in the table, the maximum population density was extremely high in Taiwan. Although the population density in combined Korea was close to that of Japan, the maximum density of Republic Korea was 517 persons per square kilometer in 2020, which was close to Taiwan. Although China contains sparse regions in inland area, the population density is as high as Taiwan in some coastal provinces.

It was calculated that the average annual growth rates in agrarian societies was 0.16% for Japan and 0.14% for Korea, which were close to Europe. The population growth during the agrarian society of Taiwan was as rapid as that in modern time because of the short length of agrarian society. The early beginning of rice cultivation in China made the annual population growth rate lower than peripheral regions. Low estimations would be given for other cradles of civilization such as Mesopotamia, Egypt, India, and Mesoamerica. The population growth after the industrialization has been incomparably faster than that in agrarian societies. Korea and Taiwan experienced especially drastic population growth with an annual growth rate higher than 1%, which implies that the population will be doubled in less than 70 years. The average annual growth rate in China has been slow because the population once collapsed in the late nineteenth century, and it was after the foundation of the People's Republic of China that the continuous population growth took place. In any case, the high population growth rates after the industrialization could not last for centuries. About 150 years passed since the modern population growth started in Japan and Korea, 100 years in Taiwan, and 70 years in China. In these countries, the population has already started declining.

After the beginning of modern era, the population grew by 3.8 times in Japan, 5.6 times in Korea, 8.8 times in Taiwan, and 3.3 times in China. The industrial revolution caused faster but shorter population growth than the neolithic revolution did. The length of modern population growth was about one and a half centuries in Eastern Asia. After populations increased by 3–9 times, the phase of long-term population decline has started. This acute feature is contrasted to slower and longer population growth in agrarian societies that increased the population by 40–90 times during several millennia. However, agrarian populations did not grow monotonously but often declined and even collapsed. A study of agrarian population decline may suggest something to the phase of long-term population decline being experienced by Eastern Asia.

3.3 Summary and Conclusion

It was supposed in the demographic transition theory that both fertility and mortality rose with the transition from a hunter-gatherer society to an agrarian society. The agriculture of China began 9500 BP and lasted for about 9300 years until the nineteenth century. Although primitive agriculture was conducted during hunter-gatherer periods in Japan, Korea, and Taiwan, this book assumed that the beginning of wet rice planting as the transition to an agrarian society. Then, Japan transitioned 2900 BP, Korea 3000 BP, and Taiwan in the seventeenth century. The agrarian society in Japan and Korea lasted for about 4800 years. During the agrarian society, the aristocrats class disappeared in the Song dynasty and the Chinese style bureaucracy with an emperor at the top was established. The Yi Joseon dynasty thoroughly imitated the Chinese institutions and introduced the agrarian bureaucracy. On the other hand, Japan transitioned to a more decentralized and competitive feudal society when the samurai warrior class took the power in the end of the twelfth century. The transition from an ancient agricultural production with a large unit to a peasant society in which an independent small family bared the production took place in the fifteenth century in China, in the seventeenth and eighteenth centuries in Japan and Korea, and in the nineteenth century in Taiwan. During the agrarian society, the population increased by 89 times in Japan, 74 times in Taiwan, and 63 times in Korea. Although the factor was 45 in China, coastal provinces showed much higher growth ratios than Japan.

Eastern Asian countries transitioned to industrial societies in the nineteenth century by the western impact. Japan quickly introduced institutions and infrastructures for a modern nation and successfully modernized under the Meiji government. On the other hand, the Qing dynasty lasted for 70 years after the First Opium War, and China exploited by the western powers. Japan became an imperial power by colonizing Taiwan and Korea.

The international order in Eastern Asia changed drastically after the end of WWII. Japan lost colonies and two nations were founded in the Korean peninsula. China was also divided when the Republic of China retreated to Taiwan and the People's Republic of China was built in the mainland. While these events distinguish the contemporary era from the modern era, there was no important change in the industrial mode of production. Demographically, however, Eastern Asian societies have recently experienced an essential transition from population growth to decline. The population growth after the modernization was faster in Korea and Taiwan than in Japan and China. This implies that Korea and Taiwan had larger room for development at the beginning of modern period. On the other hand, the start of population decline in Eastern Asia implies that a highly developed industrial society cannot sustain its population without accepting immigrants from less developed societies.

References

Bellwood, Peter. 2005. Examining the farming/language dispersal hypothesis in the East Asian context. In *The Peopling of East Asia: Putting Together Archaeology, Linguistics and Genetics*, ed. Laurent Sagart, Roger Blench, and Alicia Sanchez-Mazas. Routledge Curzon.

Caldwell, John C. 2006. The analytical approach, In *Demographic Transition Theory*, ed. John C. Caldwell, 3–19. Springer.

Cumings, Bruce. 2005. *Korea's Place in the Sun: A Modern History*, updated ed. W. W. Norton & Co.

Diamond, Jared. 1998. In search of Japanese roots. *Discover Magazine*, June 1, 1998.

Eisenstadt, S.N. 1996. *Japanese Civilization: A Comparative View*. University of Chicago Press.

Fujio, Shin'ichiro. 2015. *Yayoi Jidai-no Rekishi*. Kodansha (in Japanese).

Goode, William J. 1963. *World Revolution and Family Patterns*. The Free Press of Glencoe.

Harari, Yuval Noah. 2011. *Sapiens: A Brief History of Humankind*. Harvill Secker.

Hayami, Akira. 2015. *Japan's Industrious Revolution: Economic and Social Transformation in the Early Modern Period*. Springer.

Ho, Ping-Ti. 1959. *Studies on the Population of China, 1368–1953*. Harvard University Press.

Kito, Hiroshi. 2007. *Zusetsu: Jinko-de Miru Nihonshi – Jomon Jidai–kara Kinmirai Shakai-made*. PHP Interface (in Japanese).

Kobayashi, Toshio. 2022. *Yamataikoku Saiko – Jookoku, Yamataikoku, Yamato Seiken*. Chikuma Sinsho (in Japanese).

Lee, Yong-Hoon. 2020. *Hannichi Shuzoku Shugi-tono Toso*. Bungei Shunju (in Japanese).

Livi-Bacci, Massimo. 2012. *A Concise History of World Population*, 5th ed. Wiley.

Miyajima, Hiroshi. 1994. Higashi Asia shono shakai-no keisei. In *Choki Shakai Hendo: Asia-kara Kangaeru*, ed. Mizoguchi Yuzo, Hamashita Takeshi, Hiraishi Naoaki, and Miyajima Hiroshi, 67–96. Tokyo University Press (in Japanese).

Mizoguchi, Toshiyuki. 2008. *Asia Choki Keizai Tokei 1 Taiwan*. Toyo Keizai Shimposha (in Japanese).

Nakamura, Satoru. 2004. Shono keiei-no kikakushiteki kento. In *Nihon Shihonshugi-to Chosen,Taiwan – Teikokushugika-no Keizai Hendo*, ed. Hori Kazuo, and Nakamura Satoru, 77–93. Kyoto University Press (in Japanese).

Okamoto, Takashi. 2013a. Chugoku-no keizai-to rekishi. In *Chugoku Keizaishi*, ed. Okamoto Takashi, 177–236. Nagoya University Press (in Japanese).

Okamoto, Takashi. 2013b. Minshin – dento keizai-no keisei-to hensen (15–19 Seiki). In *Chugoku Keizaishi*, ed. Okamoto Takashi, 177–236. Nagoya University Press (in Japanese).

Shikata, Hiroshi. 1976. *Chosen Shakai Keizaishi Kenkyu Chu*. Kokusho Kankokai (in Japanese).

Starosta, Stanley. 2005. Proto-East Asian and the origin and dispersal of the languages of East and Southeast Asia and the Pacific. In *The Peopling of East Asia: Putting Together Archaeology, Linguistics and Genetics*, ed. Laurent Sagart, Roger Blench, and Alicia Sanchez-Mazas. Routledge Curzon.

Tetsuyama, Hiroshi. 2000. Chugoku dento nogyo keizai-to sono kindaiteki henyo. In *Higashi Asia Keizaishi-no Shomondai*, ed. Matsuda Koichi, 17–37. Aunsha (in Japanese).

Ueda, Makoto. 1995. Shinchoki chugoku-no saigai-to jinko. In *Koza Bunmei-to Kankyo 7. Jinko, Sippei, Saigai*, ed. Hayami Akira, and Machida Hiroshi, 236–248. Asakura Shoten (in Japanese).

Umezao, Tadao. 1967. *Bunmei-no Seitai Shikan*. Chuo Koronsha (in Japanese).

Wittfogel, K.A. 1959. *Oriental Despotism*. Yale University Press.

Zhao, Wentan, and Xie, Shujun. 1988. *Zhongguo Renkoushi*. Renmin Chubanshe (in Chinese).

Chapter 4
Population in Premodern Eastern Asia

4.1 Population and Agriculture

Although populations in Eastern Asia grew by 40–90 times during the agrarian era that lasted for millennia, the quantity and quality of food per capita declined Harari (2011). The dependence on a specific crop not only worsened the nutritional balance but also increased the risk of famine. However, the increase in mortality was compensated by high fertility, and the population and agricultural production eventually grew by dozens of times. An expansion of cultivated area or a rise in productivity caused a population increase, and a population growth for some reason stimulated an increase in production. Although it is difficult to determine the causal relationship between a population and agricultural production, cases of simultaneous growth for a long-term can be found in the history of Eastern Asia.

The principle of population proclaimed by Malthus (1798) was that population growth is checked by the limit of food production. He emphasized the inelasticity of agricultural production to population, saying that "Population, when unchecked, increases in a geometrical ratio. Subsistence increases only in an arithmetical ratio." If a population is not checked by a war, epidemics, or other disaster, it reaches at the maximum level that the existing production system can support. In this Multhusian equilibrium, the standard of living falls to the subsistence level on which the people manage to survive with the minimum amount of food. If a population exceeds this line, a famine hits the society, and the population size goes back to the equilibrium.

Boserup (1965) criticized the assumption of inelasticity and asserted that, in a long-term, a population growth can induce a growth of agricultural production through an intensification of land use. When a population was sparse, early farmers lived on fire-fallow cultivation. In such a primitive agriculture, the land after the cropping was left for decades so that a forest regenerated. If a population grew, the land was used again when the regeneration phase was still a bush. A further population increase induced a further intensification of land use, and the mode of

T. Suzuki, *Comparative Population History of Eastern Asia*,
https://doi.org/10.1007/978-981-99-9367-3_4

cultivation transitioned from short-fallow cultivation to annual cropping, and even to multi-cropping.

The history of agriculture in Eastern Asia also shows such evolution. In the case of Japan, the Jomon people were practicing the slash-and-burn agriculture. The wet rice cultivation did not spread out from northern Kyushu for centuries after the Yayoi people settled there (Fujio 2015). Probably the population pressure was not large enough to motivate out-migration and land reclamation at that time. Even after the wet rice cultivation spread in a large part of Japan, it was difficult to maintain paddies and practice annual cropping. It was estimated that, in medieval Japan, more than a half of wet rice paddies were left as fallow fields (Sato 2002). The multi-cropping was already started by the thirteenth century in the area surrounding Kyoto, but it did not spread in other areas where the population was not dense enough (Farris 2006).

Boserup (1965) discussed that while the change in the kind of tool such as from digging stick to hoe, spade, and plough corresponded with the intensification of land use, the change in the make of tool such as from wood to bronze and iron did not matter. Iron tools of agriculture began to be used from the Spring and Autumn period (770–476 BC) in China (Okuno 2000) and from the third century BC in Japan (Fujio 2015). If Boserup was correct, such changes did not induce a notable change in the population and agricultural production.

Although there were moderate population increases in the ancient era, it was only after the twelfth century that prominent simultaneous growths of the population and the production took place. For example, the population of China could not excess the level of seventy-two million that recorded in the second century for a long time. It was in the twelfth century that the Chinese population exceeded a hundred million. It took for seven centuries that the population of Japan increased from six million to twelve million in the sixteenth century. Although little is known about the population of premodern Korea, it was probably in the seventeenth and eighteenth centuries that the population grew significantly with the spread of the multi-cropping. It is difficult to determine the transition from a fallow cultivation to an annual cropping in Eastern Asia. The situation of medieval Japan can be seen as a kind of fallow cultivation. If the situation was usual in Eastern Asia, many regions transitioned directly from a fallow cultivation to a multi-cropping.

4.2 Positive Checks

A population growth is checked either via a rise in mortality or a fall in fertility. The former undesirable way of control is called the "positive check," while Malthus recommended the "preventive check" that reduces fertility through late marriage. On the other hand, the reduction of marital fertility using contraception and sterilization is called "neo-Malthusian" method. As for positive checks, Malthus thought that a war, an epidemic, and a famine were induced in this order by a population growth. It should be noted, however, that these disasters can happen independently from the population pressure. Although a food shortage caused by the population pressure can result in a riot, it is rare to induce a large-scale war. Harari (2011) pointed out that WWII was not caused by a food shortage and warned historians in the future not to give an easy demographic or economic explanation. A war can happen to reduce the excess military power and to reward subordinate generals, as in expeditions of Alexander the Great, Genghis Khan, and Toyotomi Hideyoshi. Since the nineteenth century, wars were motivated by the imperialism that attempted to extend national power and domination. Needless to say, a religious or ideological conflict can result in a war.

The transition from a hunter-gatherer society to an agrarian society not only raised the risk of infection through the dense inhabitation but also brought about new epidemics from domesticated animals. Measles and smallpox were originally diseases of cattle, and influenza infected from ducks to humans (Diamond 1997). According to McNiell (1976), an agrarian population increased and decreased in the interaction between disease orgasm and human hosts. A new lethal disease can kill a huge number of people but calms down as the people acquire the immunity, and finally turns into a childhood disease that everyone experiences. However, once the lethal disease is brought into a virgin land, a disastrous outbreak occurs and causes a massive number of deaths. The pandemic of the plague in Europe and western Asia in the fourteenth century broke out when the ecological balance was destroyed by the foundation of the Mongol empire. The European conquest of America caused a catastrophic collapse of the Amerindian population mainly due to measles and smallpox. According to McNiell's view, most infectious diseases in Eurasia were verged toward the harmless status by the eighteenth century owing to the acquired immunity, preparing the continuous population growth. China was the earliest to catch this opportunity, followed by the UK under the industrialization.

The plague pandemics killed one quarter of the Chinese population in the seventh century and one third of the European population in the fourteenth century (Ishi 1995). Although there remains no record on the plague in premodern Japan or Korea, the possibility of international transmission will be examined in this chapter. Virus diseases such as measles, smallpox, and influenza were well recorded and had considerable demographic impacts. While the death toll by measles and smallpox decreased in the nineteenth century, tuberculosis became a major cause of death. As in other parts of the world, Eastern Asia could not escape from the global pandemics of cholera (1817–23), Spanish flu (1918–20), and COVID-19 (2020–22).

Famines occurred independently from the population pressure, as wars and epidemics did. A warm weather resulted in droughts, while a cold summer also caused a bad harvest. It seems that a cool weather had larger demographic impacts through great famines than a warm weather. China experienced drastic population collapses due to the cool weather in the third and fourteenth centuries (Okamoto 2013a, b). In Japan, a cool weather induced more serious famine than a warm weather (Kikuchi 2000). Two greatest famines of Korea in 1670–71 and 1695–96 were also caused by the cold summer. It is hypothesized that the cool weather in the thirteenth century was caused by the catastrophic eruption of the Samalas volcano in Indonesia in 1257 (Tange 2014). Less catastrophic eruptions could cause serious damages through cold summers and great famines. In Eastern Asia, explosions of the Mt. Paektu (969 ± 20), the Mt. Fuji (1707), and the Mt. Tarumae (1739) had large impacts.

A famine can be caused by a war or a rebellion. The Mongol invasion to Korea in the thirteenth century and the rebellions in the nineteenth-century China especially had drastic demographic damages. In the twentieth century, some famines were caused by the governmental failure. The Bengal famine of 1943 occurred because the British government ordered to export a huge amount of foods without considering the food supply in the colony. The Great Leap Forward policy (1955–62) led by Mao Zedong resulted in a disastrous famine. The DPK Korea government could not cope with the end of the Cold War regime and the people suffered from a great famine.

4.3 Climate Changes

The global climate change had a significant impact on the agricultural production and the population dynamics. Especially, a cool weather not only increased the damage from famines but also stimulated large scale migration and invasion wars. Table 4.1 summarizes the climate changes in the past three millenniums. The beginning of the ancient cool period is ambiguous. It is thought that Aryan migration from Iran to India was motivated by the cool weather around 1500 BC (Kazama 1993). Both Shang dynasty and Zhou dynasty were founded by barbarians invaded Middlefield to avoid the cool weather (Okuno 2000).

After the warm period for four centuries, the climate cooled again in the third century. The population of China decreased because of the cool weather in the third and fourth centuries (Okamoto 2013a). Paleoclimatologists theorize that there was a large volcano eruption, resulting in extreme weather events in 535. China was hit by horrible droughts in 536–38 (Fagan 2004). The increase in migration into Japan after the third century can also be attributed to the cool weather (Tange 2014).

The medieval warm period lasted from the eighth to the fourteenth century. In China, there were frequent droughts in the north and floods in the south. While western Japan suffered from droughts, eastern Japan enjoyed good crops because damages from cool summers decreased (Tange 2014). In Japan, the period between

Table 4.1 History of climate changes

Period	Climate
-200 BC	Ancient cool period
200 BC–200 AD	Ancient warm period
200–700	Medieval cool period
700–1300	Medieval warm period
1300–1800	Little ice age
1800–	Modern warm period

the twentieth century and the early thirteenth century was the peak of warm and dry weather. The great famine in 1181 was caused by a drought (Kito 1995).

The warm period was succeeded by the little ice age between the fourteenth and nineteenth centuries. As mentioned above, the cool weather was possibly started by an eruption of volcano. The population of China declined because of cool weather in the fourteenth century (Okamoto 2013b). Great famines in Korea occurred in the seventeenth century when the cool weather was at the peak (Kim 2010a). In Japan, not only the frequency of famine but also the risk of flood rose in the fifteenth century (Mizuno 2006). The damage of the great famine in 1459–61 was especially serious (Minegishi 2011). During the Edo period, three major famines in 1732–33, 1783–88, and 1833–39 happened because of the cool summer.

In the current period of global warming, the process is assumed to be accelerated by the emission of greenhouse gases through the human activity. Anticipated risks of the global warming include submergence of coastal cities by sea level rising, an increase in storms and droughts, mass extinction of species, a collapse of agricultural and fishery production, an expansion of tropical diseases, and so forth. However, the global population growth has been slowed, and the technology of energy saving and substitution has been developed. The international consensus has been made, even though too slow to satisfy the experts. Damages from the global warming may not be as severe as those in the little ice age.

4.4 Major Population Collapses

Many societies experienced drastic population collapses, and even extinctions that were not recorded. In fact, a large part of the agrarian period was not recorded in most countries. For China, the historical records cover at most one third of its agrarian period that lasted for almost ten millenniums. In the case of Japan, own historical records written by the Japanese cover less than a half of the agrarian era. There should have been population disasters that were once recorded but were lost today. It is known that *Samguk Sagi*, the oldest remaining historical book written by Koreans, was heavily edited according to Confucian values. There should be many historical events in the original sources that were not included in the book.

Among recorded population collapses outside of Eastern Asia, three cases are famous for disastrousness. According to McNeill (1976), Eurasian peoples had achieved the epidemiological balance with various infectious diseases by the tenth century. This balance was broken by the foundation of the Mongol empire in the thirteenth century. A communications network of post messengers, commercial caravans, and armies was built, and wild rodents of the steppe lands came into touch with carriers of new diseases including bubonic plague. Although the plague was absent in Europe for centuries since 764, the epidemic proceeded from Crimea to penetrate almost all of Europe in 1347. It is supposed that the pandemic referred to as the Back Death killed one third of the population in the fourteenth-century Europe.

The conquest of the Aztec empire by Hernán Cortés (1485–1547) and that of Inca empire by Gonzalo Pizarro (1502–48) succeeded because of the virgin land effect of measles and smallpox. Eurasian diseases such as typhus, diphtheria, and mumps hit the New World one after another. Malaria and yellow fever prevailed in tropical and subtropical America (McNeill 1976). Estimates of indigenous American populations cited by Livi-Bacci (2012) suggests that 90% or more were lost after the European conquest in the sixteenth century. According to him, although the main cause of this terrible population decay was the virgin land effect of Eurasian and African diseases, other factors such as genocide, forced labor, and the reorganization of production also worked.

The Irish Potato Famine (1845–52) that brought about a disastrous collapse of the Irish population was caused not by a disease of human but by that of potato. In 1845, a fungus, *Phytophthora infestans*, destroyed the potato harvest of Ireland where people were dependent on that food. A great famine hit the island and epidemics of fevers and typhus prevailed in 1846–47. Because absentee landlords and food merchants in England lobbied against the export ban, food exports from Ireland continued during the famine. The number of excess deaths exceeded one million, and those who managed to survive left the island. With excess mortality and emigration, the population of Ireland almost halved in 1841–1901 (Livi-Bacci 2012).

China experienced several population collapses. The Chinese population decayed drastically after the peak at the mid-second century. McNeill (1976) attributed it to the epidemic of new diseases. According to him, three or four out of ten men died for the 162–63 pestilence. Infectious diseases with severe die-offs recurrently visited China for centuries. The great pestilence in 310–12 left only one or two out of a hundred persons alive in northwestern provinces of China. These cases suggest the arrival of hitherto unknown infection, supposed to be measles or smallpox, or both. Okamoto (2013a, 2015) discussed that the population decay was caused by the decreasing agricultural production owing to the global cooling. As mentioned above, the medieval cool period began in the third century. It was usual that the Chinese population decreased in the end of a dynasty. In the case of the Han, a nationwide religious revolt called the "Yellow Turban Rebellion" broke out in 184, causing massive excess deaths and disturbing the agricultural production. Although the imperial army suppressed the rebellion within a year, smaller riots recurrently happened, and the socioeconomic disorder continued until the Han dynasty finally fell in 220.

After the population crisis in the end of the Han dynasty, the Chinese population stagnated for centuries and could not recover until the Tang dynasty. On the other hand, acute population decreases in the Qin and Sui dynasties swiftly compensated in the successive dynasties. The Qin and the Sui lasted only for a short period and replaced by the Han and the Tang, respectively. When Qin Shi Huang, the first emperor of China, died in 210 BC, the first peasants' uprising in Chinese history led by Chen Sheng and Wu Guang broke out. The battles between the Qin and various rebellion forces continued until the Qin eventually ruined in 206 BC. A legend says that the inhabitants of Jin Han in southeastern Korea in the third century were the offspring of refugees escaped from the Qin's rule. This suggests that there was a large size of exodus, not necessarily to Korea, in the end of the Qin period. The fall of the Sui dynasty was caused by the ambitious enterprises of the second emperor Yang (reign 604–618) such as the construction of the Grand Canal and the campaign against Goguryeo. After the rebellion of Yang Xuangan in 613, dozens of rebel leaders occupied significant portions of the dynasty and fought each other until the Tang eventually replaced the Sui in 618, causing severe economic and demographic damages.

The Chinese population in the ninth century regained the size in the first century but could not sustain it. The population decreased in the late Tang period and stagnated during the Five Dynasties and Ten Kingdoms period. The rapid population growth caused by the agricultural revolution in the Song period was disrupted by the successive invasions of the Liao, the Jin, and the Mongol empire. McNeill (1976) attributed the reason of the population collapse in the fourteenth century to the recurrent epidemics of plague due to the broken epidemiological balance. In 1331, the plague prevailed in Henan province and killed nine tenth of the population. A disastrous period of plague coincided with civil war continued until alien rulers were overthrown and the Ming dynasty was established in 1368. However, the impact of declining agricultural production due to the global cooling should not be ignored because the little ice age had started in the fourteenth century. In fact, the Chinese economy was at the bottom of depression in the beginning of the Ming dynasty (Okamoto 2013b, 2015).

Ho (1959) emphasized the severity of population collapse in the seventeenth century when the Qing dynasty replaced the Ming. According to him, the widespread peasant rebellions drastically reduced the populations of many localities in north China and nearly exterminated the population of Sichuan basin. The area of officially registered land shrank to almost one-half, and the flow of migrants to Sichuan province lasted for two centuries. Maddison (2001) wrote that twenty million people died owing to wars, famines, and smallpox in this period. The maritime bans in the early Qing period depressed the economic and population growth (Okamoto 2013b). According to Saito (2002), however, the population decay in the seventeenth century was exaggerated by the incompleteness of statistics and was not as severe as that in the fourteenth century.

It was measles or smallpox, or both, that collapsed the Han population in the second century (McNeill 1976; Ishi 2018). It is natural to guess that these eruptional diseases were transmitted to ancient Korea and Japan to result in large population disasters. However, the period was too old to be studied with own historical records in Korea or Japan. The oldest case of eruptive epidemic in Japan was that in 552. The disease was called "inamegasa," suggesting that rashes looked like grains of rice appeared (Fujikawa 1912). It is more difficult to identify pathogens in Korean history. The oldest case of smallpox was that in the eighth century, which was transmitted to Japan and caused the pestilences in 735 and 737 (KSID 2009). In those historical periods, however, Korea and Japan were not virgin lands anymore and the demographic impacts may not have been as severe as that in the second-century China.

The disaster of plague that destroyed the population of Europe and China in the fourteenth century cannot be found in historical records of Japan or Korea. Maddison (2001) discussed that Japan was isolated from the continent and was free from the plague. According to Yasuda (1995), the ecological environment of Japan with irrigated rice paddies and a lot of natural enemies of rodents prevented the pestilence of plague. However, McNeill (1976) pointed out the possibility that the disease in 808 was plague. According to *Goryeosa*, roads of Kaesong, the capital, were covered with corpses due to famine and epidemic in 1348. This disease could have been plague, although its demographic impact is unclear. Plague prevailed in China at the end of the Ming dynasty. The possibility of transmission of plague from China to Korea and Japan at that time will be discussed later. In any case, an impact of plague as serious as in Europe cannot be confirmed in Korea and Japan.

Korea experienced the population crisis due to the Mongol invasion in the thirteenth century. The Mongol empire invaded Korea nine times in 1231–59. The royal court of Goryeo escaped to Ganghwa island, and the military rulers began a thorough resistance. In the third invasion in 1234–38, the Mongols burned farmlands to disrupt the logistics. In the sixth campaign in 1253–58, the death toll was uncountable, bodies covered the land, and the whole country was destroyed. Although the royal court surrendered in 1259, the resistance of Korean guerrilla lasted until 1273.

The emperor Khublai (reign 1264–94) decided to beat Japan and ordered the Goryeo palace in 1274 to build 900 battleships and mobilize 35,000 Korean men. The burden seriously disturbed the agricultural production, and most Korean men went to Japan could not return. Khublai forced the Goryeo again to provide a huge amount of human and material resources for the expedition to Japan in 1281. The Goryeo dynasty suffered unrecoverable damages from nine invasions and two campaigns. Kim (2010b) estimated that the population of Goryeo decreased by 46% in 1231–81.

Although the Yi Joseon dynasty underwent the Japanese invasion (1592–98), the Manchu invasions (1627, 1636–37), and great famines (1670–71, 1695–99), demographic impacts should have been smaller than that in the thirteenth century. Most historians agree that the population decreased in the first half of the nineteenth century, although its intensity was milder than foregoing population collapses. Furthermore, Cha (2014) asserted that the Korean population did not decline in that period.

As for Japan, no drastic population collapse can be found except for possible epidemic disasters in the prehistoric period. According to Farris (2006), impacts of epidemics and famines reduced in the medieval period, and excess mortality from battles during the warring state era was smaller than that in the American civil war. Although the Japanese population stagnated in the eighteenth century, there was no population collapse during the Edo period. Three major famines in 1732–33, 1783–88, and 1833–39 had smaller demographic impact than those in the seventeenth-century Korea.

4.5 Premodern Population Estimates of China

Table 4.2 summarizes population of China until the seventeenth century. Kobayashi (1942) collected population sizes recoded in history books without any adjustment. Maddison (2001) attempted backward projections of population and GDP with assumed growth rates for each country. Estimates of Zhao and Chen (1989) were cited in Tetsuyama (2000), and those of the National Population and Family Planning Commission (2011) were cited in Wakabayashi and Nie (2012). The Chinese population was already so huge in the ancient time that estimates are shown in million, not in thousand.

In the waring state period of China, the Qin kingdom became increasingly powerful under reforms led by Shang Yang (390–338 BC). After the thirty-first king Zheng, later the Qin Shi Huang, took the throne in 247 BC, the Qin kingdom overthrew rivals one after another, and finally unified northern and southern China in 221 BC. However, the Qin dynasty fell soon after the death of Qin Shi Huang. The population decayed during the confusion between the first peasants' rebellion (209 BC) and the foundation of the Han dynasty (206 BC). The main cause of the rebellion was the harsh law enforcement. As suggested by the legend of Jin Han in Korea, there should have been a large amount of diaspora during the Qin period. According to the estimates by the NPFPC, the Chinese population fell from thirty million in the middle of the waring state period to twenty million at the unification by the Qin. The population decreased further to thirteen million shortly after the reunification by the Han. This exemplifies how China spent power for the unification.

The oldest remaining population register was compiled in 2 AD and counted about sixty million. It is possible that the Han population grew further to be seventy million or more until disrupted by the epidemics of measles and smallpox in the late second century. Although the total population in 156–57 recorded in the *Book of the Later Han* was about fifty million, the NPFPC estimated that the actual population was seventy-two million. The population collapsed owing to the eruptive diseases and the cool weather to be, according to the NPFPC, twenty-five million in the third century. The population stagnated throughout the ages of ununified China, namely the periods of three kingdoms (220–280), sixteen kingdoms of the five barbarians (304–439), and northern and southern dynasties (439–589). Despite the recurrent flows of five barbarians into the Middlefield, the population could not grow uninterruptedly because of the frequent epidemics and the deteriorated agricultural production.

The second emperor Yang of the Sui dynasty mobilized a huge amount of labor forces for the construction of the Grand Canal, which completed in 610. *Cefu Yuangui*, an encyclopedia edited in the early eleventh century, recorded that the population at Yang's early reign was forty-six million. However, the NPFPC estimated that the actual population was fifty-five million. After the bruising battles among dozens of rebellion leaders, Li Yuan (566–635) forced the last emperor of the Sui to transfer the throne and founded the Tang dynasty in 618. However, his reign (618–26) was spent to wipe out the remaining rebel groups. According to the NPFPC, the population

Table 4.2 Population estimates of China until seventeenth century (million)

Year	Kobayashi (1942)	Maddison (2001)	Zhao and Chen (1989)	NPFPC (2011)
340 BC				30
221 BC				20
202 BC				13
2	60	59	59	63
57	21			
105			53	53
146			47	47
156–157	50			72
265				25
300				33.79
368–407				31.28
520				52.4
581				44.3
606	46			55.42
624				22.74
755	53			87.75
860				67
960			32	39.79
1000		59		
1072	14		121	
1110	47			119.46
1160	19			
1190	59			
1207–23				125.4
1351				97.3
1393	61		60	
1500		103		
1566				164.8
1578–81	61		200	
1600			160	
1644–61	120			
1662			83	

collapsed to be less than a half because of the confusion caused by the emperor Yang.

The Tang dynasty was at the peak during the reign of the seventh emperor Xuanzong (reign 712–56). However, the emperor lost attentions to political matters as he sought the pleasure with his concubine Yan Guifei, and over-trusted An Lushan (703–57) who led the rebellion in 755–63. *Tongzhi*, an encyclopedia edited in the twelfth century, recorded that the population immediately before the An Lushan rebellion was fifty-three million. According to the NPFPC, however, the estimated population was about eighty-eight million, which updated the peak during the Han period after the long stagnation for seven centuries. The rebellion of an seriously damaged the economic production, and the Tang could not sustain the population. Estimates contained in the NPFPS (2011) show that the Tang population declined by 24% in 105 years. The population declined further during the period of five dynasties and ten kingdoms (907–60).

When Zhao Kuangyin (927–76) built the Song dynasty in 960, the population was less than a half of the peak in the Tang dynasty. It is believed that the population excessed a hundred million during the Northern Song period (960–1127). Kobayashi (1942) pointed out that the population register of the Song implies the average household size of only two or less, and that the number of household times five gives a population of about one hundred million. Zhao and Chen (1989) estimated that the population exceeded one hundred million in the late eleventh century. The massive population growth in the eleventh century was achieved by the governmental effort to develop southern provinces. The third emperor Zhenzong (reign 997–1022) introduced the Champa rice, an early-ripening and drought-resistant variant, to lower Changjiang and Huai River regions. The early-ripening characteristic made it possible to use the same land to wheat, rape, or other winter crops after the rice was harvested. Since the Champa rice required much less water than other varieties, higher land and hilly slopes were newly cultivated, which also helped the dramatic growth of the food production (Ho 1959). The simultaneous growth of agriculture and population stimulated manufacturing and service industries, resulting in the first commercial revolution in China (Okamoto 2015). McNeill (1976) emphasized the epidemic factor, writing that an acquirement of human resistance to malaria, bilharzia, and dengue fever was one of preconditions of the massive population growth.

The Northern Song fell when the Jin army occupied the capital and captivated the emperor Qinzong in 1126. His brother, Zhao Gou, fled to Nanjing and established the Southern Song (1127–79). The Jin dominated northern China until the Mongol empire replaced it in 1234. Khublai Khan declared the foundation of the Yuan dynasty in 1271 and demolished the Southern Song in 1279. The rule of the Yuan ended when Zhu Yuanzhang, a rebel leader, built the Ming dynasty in 1368. According to the NPFPC (2011), the population reached 125 million in the early thirteenth century when the Jin and the Southern Song divided China. However, the population declined to 97 million in 1351 when Zhu Yuanzhang participated in the White Lotus rebellion. The population decayed further to 60 million in 1393, which was less than a half of the peak in the Song period. As discussed above, the collapse was caused by the epidemic of the plague and the agricultural failure due to the cool weather. Zhao and

Chen (1989) assumed that the recorded population of about 60 million in 1393 was correct, suggesting that the population continued to decline in the early Ming period. The economy gradually recovered in the fifteenth century owing to the development of the Changjiang middle basin as the rice production area. While Zhao and Chen assumed that the Chinese population barely regained one hundred million in 1500, Maddison's estimation implies a higher growth rate.

The population growth in the sixteenth-century China was stimulated by the establishment of the global trade system by Europeans. The Portuguese constructed a basis in Macau in 1557 and began trades with Chinese and Japanese pirates. The Spanish occupied Philippines by 1571 and traded Chinese silk and porcelain with Mexican silver between Manila and Acapulco. The Dutch established trading posts of Java and Japan in 1603 and 1609, respectively. The British retreated from Eastern Asia after the competition with the Dutch (Maddison 2001). Peanuts, sweet potato, and maize were introduced to China in the sixteenth century and Irish potato in the seventeenth century. Uncultivated dry hills and mountains gradually turned into maize and sweet potato farms. As rice culture approached to its limit, American dryland food crops contributed to the increase in national food production and made possible a continual population growth (Ho 1959). It is unrealistic to conclude that the population did not grow in the sixteenth century, as suggested by the historical records that Kobayashi referred to. The NPFPC estimated that the population was 165 million in 1566, and Zhao and Chen suggested that the population reached 200 million in 1578–81.

As money uses and commercial transactions spread nationwide by the late sixteenth century, the wealth and the land ownership more and more concentrated on scholar-gentlemen who passed the imperial examination (Okamoto 2013b). The increasing inequality and poverty induced peasant rebellions, and the dynasty had financial difficulties to cope with the domestic upheaval in addition to the war against Japan in the Korean peninsula. The Ming fell when the army led by Li Zicheng occupied Beijing in 1644, and the Qing's army immediately replaced Li and declared the rule of China. Although Saito (2002) was skeptical, Zhao and Chen judged that the population collapse in the seventeenth century was as catastrophic as that in the fourteenth century. According to their estimates, the Chinese population declined from 200 million in 1581 to 83 million in 1662. On the other hand, Kobayashi concluded that the population during the reign of Shunzi emperor (1644–61) was as large as 120 million. He analyzed the population register reports contained in a historical book entitled *Donghualu*. The book contains only the number of adult males until the 1730s. Since the adult male population was about 14 million during the Shunzi period, he assumed the ratio of adult male to total population was about 12%, which seems too small if the population register was complete. He guessed that the number was underreported because an adult male was the unit of taxation.

Table 4.3 compares the population estimates for the Qing dynasty before the First Opium War. Estimates by Kobayashi (1942) and Zhao and Chen (1989) were included again to examine the population growth in the early Qing period. Kobayashi deduced from the adult male population that the total population increased from 130–40 million to 170–80 million during the Kangxi period (1662–1722). While Chen (1979) estimated that the Chinese population was 150 million in 1700, Maddison

(2001) assumed a smaller population. Although *Donghualu* recorded that the total population in 1742–45 was 160–70 million, Kobayashi asserted that the reality was around 200 million. Both Kobayashi and Chen concluded that the numbers in 1775–94 were acceptable without adjustment. Zhao and Chen (1989) guessed that the population in 1785 was smaller than the recorded population of 289 million. For the first year of the Daoguang emperor (reign 1820–50), while Zhao and Chen assumed again that the population was smaller than the record, both Cao (2001) and Maddision (2001) concluded that the population was larger.

If we accept recorded populations in *Tonghualu*, as Kobayashi and Chen did, the population growth was already fast in the early eighteenth century. The growth from 150 million in 1700 to 200 million in 1745 implies the annual growth rate of 0.64%. The growth accelerated to be 0.92% in 1745–94. The Qing dynasty was in the golden age during the reigns of the Knagxi emperor (reign 1661–1722), the Yongzheng emperor (reign 1722–35), and the Quanlong emperor (reign 1735–96). During the Kangxi period, the increased international trade led the economic boom. China exported raw silk, tea, sugar, porcelain and other manufactural products, and imported silver from Europe, rice from Thailand, and cotton from India. Especially, the tea exportation grew acutely owing to the growing demand in Britain, and a huge amount of silver flew into China. Although not shown in Table 4.3, Okamoto (2013b) estimated that the Chinese population reached 300 million in the mid-eighteenth century, and 400 million in the early nineteenth century. Then, it is difficult to accept the conclusion of Zhao and Chen that the population stagnated in 1785–1820. The rapid population growth was supported by the expansion of agricultural fields. Dry mountainous areas in Jiangxi, Hubei, Hunan, and Sichuan provinces turned into maize and sweet potato fields. When the cultivation of these provinces reached a limit, peasants migrated to Yunnan, Guangxi, Taiwan, and northeastern provinces and expanded fields (Okamoto 2013b).

Table 4.3 Population estimates of premodern Qing (million)

Year	Kobayashi (1942)	Chen (1979)	Zhao an Chen (1989)	Cao (2001)	Maddison (2001)
1644–61	120				
1662	130–40		83		
1700		150			138
1722	160–70				
1745	200				
1780	278	275			
1785	289		268		
1790	301				
1794	313	313			
1820	356		295	383	381

According to McNeill (1976), Eurasian societies were ready for population growth by 1700. The impact of Mongol empire that destroyed the epidemiological balance was gradually overwhelmed, and a new epidemiological balance was achieved by the end of the seventeenth century. Most infectious diseases became childhood diseases that everyone contracted, leading people acquire immunity to them. McNeill asserted that China was the first region to take the opportunity. While parallel potentialities elsewhere were masked by a variety of countervailing circumstances, public peace prevailed unbroken and conventional limitation upon taxes and rents remained well defined. The population growth of England and Wales started in the late eighteenth century, which was half a century later than China.

4.6 Premodern Population Estimates of Japan

Table 4.4 shows population estimates of Japan between the Nara and Edo periods. Estimates with intervals for 1280, 1450, and 1600 were by Farris (2006), and all other figures were taken from Table 1 of Kito (2007). The outline of the Jomon population changes was discussed above and not included here. There is no estimate for the Yayoi, Kofun, and Asuka periods. There could have been catastrophic population collapses due to civil wars, famines, epidemics, and other natural disasters during these periods. Especially, the first hit of a new disease must have had a disastrous impact. Considering incessant migration flows from the continent, lethal infectious diseases could have transmitted to Japan soon after the emergence in China.

The importance of immigrants in ancient Japan can be seen from *Shinsen Shojiroku*, a genealogical record edited in the ninth century. Of 1182 families, 163 families proclaimed the Chinese ancestry, and another 163 families reported the Korean ancestry. Thus, immigrants from China and Korea accounted for more than a quarter. Although there remains no record on their arrival time, it is supposed that they mostly came after the fifth century. The Yamato rulers gave preferential treatments for foreigners who had military, administrative, agricultural, and manufactural skills. Such information was transmitted to Korea and China, and induced migration to Japan. The cool weather and wars worked as the push factors in the continent. The unification war in the seventh-century Korea resulted the last big wave of migration to Japan. The migration flow mostly stopped after the Nara period (Kito 2007).

The estimates for 725 and 800 in Table 4.4 were made by Kamada (1984) and Sawada (1927), respectively, being based on the taxation records. Although the KSID (2009) wrote that the Japanese population declined by one half owing to the epidemics of smallpox in 735–37, population estimates do not support such a catastrophic incident. The estimates for 900 and 1150 were calculated from the land registers recorded in encyclopedias named *Wamyo Ruijusho* and *Shugaisho*, respectively. Even with frequent outbreaks of lethal infectious diseases in ancient Japan, the population increased by 43% in 725–900. The agricultural production in the eighth-century Japan grew owing to the use of iron farming tools and draft animals in addition to the construction of reservoirs and irrigation canals. The growth continued to the ninth

Table 4.4 Premodern population of Japan

Year	Population (thousand)	Year	Annual growth (%)
725	4512	725–800	0.27
800	5506	800–900	0.16
900	6441	900–1150	0.02
1150	6837	1150–1280	− 0.13 to − 0.08
1280	5700–6200	1280–1450	0.26–0.36
1450	9600–10,500	1450–1600	0.10–0.38
1600	12,273–17,000	1300–1721	0.50–0.78
1721	31,279	1721–1732	0.29
1732	32,306	1732–1750	− 0.23
1750	31,011	1750–1756	0.15
1756	31,283	1756–1762	− 0.09
1762	31,106	1762–1780	0.02
1780	31,213	1780–1786	− 0.60
1786	30,104	1786–1792	− 0.13
1792	29,870	1792–1798	0.38
1798	30,565	1798–1804	0.10
1804	30,746	1804–1822	0.21
1822	31,914	1822–1828	0.37
1828	32,626	1828–1864	− 0.08
1834	32,477	1834–1840	− 0.72
1840	31,102	1840–1846	0.63
1846	32,297		

ISE (1974), Kito (2007), Farris (2006)

century due to the governmental policy to induce the cultivation of new rice fields. The government also encouraged to plant wheat and barley, which motivated multi-cropping (Kimura 2010). Such an agricultural growth approached to a limit and the population growth slowed in the tenth century.

In the Western history, the medieval time is often thought as a dark and stagnated period between the ancient glory and the modern enlightenment. The population collapse owing to the Black Death can reinforce this popular view. Although little is known about the population in medieval Japan, there seems to have been no population collapse as disastrous as those in Europe and China. Farris (2006) did not see that the medieval time began with the establishment of Kamakura bakufu in 1192 but defined the period 1150–1280 as the early medieval era. The land register for 1150 recorded in *Shugaisho* was the last nationwide record, and only provincial registers called *Otabumi* remain after that year. Farris estimated provincial populations for around 1280 and found that the population declined acutely from 1150 in central and eastern Japan while increased in northern Kyushu. His estimation implies that the total population declined by 10–17% during this period. Although the population decline was less catastrophic than those in the Eurasian continent, the Japanese population stagnated throughout the early medieval period. According to Farris, the main cause was a series of great famines in 1180–82, 1229–32, and 1257–60 that brought about much higher excess mortality than famines in the Edo period.

The lower limit of the total population in 1450 is Farris' own estimation based on military records. He calculated that 132,775–197,760 soldiers were mobilized nationwide in the fifteenth century, which was 21–80% larger than Nara period. He applied this growth ratio to the rural population, and added the urban population assuming that the percentage of urban was 4% in 1450. He accepted an unpublished estimate by Saito Osamu as the upper limit of the total population. Unlike in the early medieval period, the Japanese population grew from six to ten million in 1280–1450. The population growth took place mainly in the "Muromachi Optimum" (1370–1450) opened by the third shogun Ashikaga Yoshimitsu (in office 1368–94). According to Farris, epidemic outbreaks decreased owing to the end of the war between the Northern and Southern dynasties and the continuous rise of immunity. Famines also decreased due to the warm weather in 1300–1450. Furthermore, developments in agronomy, engineering, and settlement patterns raised fertility. Innovations such as the introduction of Champa rice, the practice of double cropping, and the use of iron tools and livestock spread from the Kinai area (today's Kyoto, Osaka, and Nara prefectures) to a large part of western Japan.

The lower limit of the population in 1600, 12.3 million, is the estimation by the Institute of Social Engineering (1974) led by Hayami Akira. The population estimates for the eighteenth century are based on the national censuses conducted by the Tokugawa bakufu and supposed to be reliable. Hayami and his colleagues divided Japan into three parts, namely the central region (Kinai), the middle-developed region (today's Gifu, Aichi, Mie, Shiga, and Hyogo prefectures), and the less developed region (other prefectures). They assumed that the population of Kinai tripled in 1500–1750, that of the middle-developed region tripled in 1550–1750, and that of the less developed region tripled in 1600–1750. They attempted backward projections to obtain the populations by province in 1600 and summed up them to get the total population. However, Japanese demographers including Fujino Shozaburo, Kito Hiroshi, and Saito Osamu felt that 12.3 million is too small and proposed new estimates ranging from 14 to 19 million. While Farris (2006) concluded that the range of 15–17 million is appropriate to 1600, Kito (2007) accepted Hayami's estimate in his Table 1. Thus, the range of 12.3–17.0 million was shown in Table 4.4.

Although Japan's population continued to grow in the late medieval era (1450–1600), it is more possible that the growth rate was smaller than the mid-medieval era (1280–1450). The average annual growth rates ranged in 0.26–0.36% for the mid-medieval period and 0.10–0.38% for the late medieval period. Farris (2006) pointed out the possibility of mortality rising during the warring state and the unification periods. Syphilis was added to the list of dangerous diseases in the sixteenth century, and the improvement in transportation promoted contacts among susceptible persons. Famines increased again as the weather cooled down in the late fifteenth century.

On the other hand, Farris speculated that fertility rose, and infant mortality fell in the late medieval Japan. The waring state era was a period of substantial economic growth. In addition to the significant growth in food production, various innovations were achieved in trade, mining, architecture, and manufacturing that contributed to improve the living standards. The corporate village and stem family continued to spread, perhaps providing stronger support systems for children. The extent to which these factors compensated rising mortality is unknown. While Hayami's estimation suggests a slower population growth than in the mid-medieval era, Farris asserted that socioeconomic developments more than offset the harsh effects of the new mortality regime.

While it is agreed in Japan that the Edo period is entirely included in the early modern history of Japan, some historians also include the Azuchi-Momoyama period or even the waring state period. Although the feudalistic political system continued and the samurai class took the power in early-modern Japan, epoch-making changes that distinguish from the medieval period took place in the economic sector. The farmland area increased rapidly owing to the boom of river construction in the seventeenth century, which brought about the massive simultaneous growth of food production and population. As the expansion of farmland approached to the limit, the intensive labor was required to improve the land productivity. The transition from the ancient agriculture that exploited dependent workers to the peasant society that relied on the hard work of independent family workers proceeded under this condition. According to Miyajima (1994), most of the villages existing today were founded after the transition to peasant society. The Japanese family system "ie" as a management body expected to perpetuate through impartible inheritance was also spread in the peasant society. Miyajima discussed that, in these features, the emergence of peasant society was more epoch-making than the modernization in Eastern Asia. However, at least demographically, the modernization had a larger impact considering the rapid population growth that tripled the population within a century, the mortality decline that nearly doubled life expectancy, the epidemiological transition that totally changed the causes of deaths, and activated international migration since the nineteenth century.

Nakamura (2005) discussed that preconditions for peasant society included agricultural development, the division of labor, the organization of village or kinship group, and the advancement of rural markets. It is assumed that these conditions were available both in Western Europe and in Eastern Asia. However, some socioeconomic features existed in Western Europe and Japan were absent in China and Korea. Hayami (2015) asserted that Japan transitioned to economic society, in which the economic value was independent of other values. While Western Europe and Japan experienced "decentralized societies" in the medieval era, China and Korea could not repudiate the ancient system and suffered from fossilization. Hayami went a step further to point out the economic change that took place only in Japan. The "industrious revolution" in early modern Japan was a capital-saving and labor-intensive agricultural production, in which animal power was replaced by the hard work of household members.

It is often discussed that the Japanese family is unique not only in Eastern Asia but also worldwide. While marriage within the paternal clan was strictly prohibited in Confucian societies, adoptions were always executed within the clan. By contrast, the Japanese family was more interested in sustaining the family name and property rather than preserving the blood line. According to Nakane (1970), this feature was derived from the basic structure of Japanese society that the "frame" is a more important criterion of groups than "attribute." In other societies, including India, China, and Europe, attributes of birth or achievement are the most important principles for group formation, whose solidity is derived from the homogeneity of members. Families and kinship groups in those societies were naturally defined through blood ties and matrimonial relations. In Japan, however, a family was considered the locality of residence and management, and locally limited membership was more important than family members living apart. This is the reason why Japanese families adopted members from outside kinship groups and considered servants to be family members. Sakane (2010) defined the Japanese family system "ie" as a management group to be sustained over generations with primogeniture and took the form of a stem family. Such "ie" institutions emerged in the samurai class in the medieval era and spread to the whole strata during the early modern era. With the custom of impartible inheritance, peasant families stayed in the same location while accumulating experience and forming dense relationships within the village. Such immobile and impartial land use enabled continuous investment and promoted industriousness.

The average annual population growth rate in 1600–1750 was 0.40–0.62%, which was apparently higher than the medieval period. It is certain that the positive feedback between the population and agricultural production enabled the simultaneous growth to continue for more than a century. The early modern daimyo exerted unilateral control over their territories. In Kanto and Osaka plains, large-scale flood control and irrigation projects were implemented by the shogunate, converting fertile river deltas to productive rice fields. Small-scale projects were conducted by village communities or wealthy farmers in various areas. As a result, seventeenth-century reclamation progressed at a pace never seen before (Hayami 2015).

In the eighteenth century, however, the pace of land reclamation decelerated, and the population growth also stagnated. As shown in Table 4.4, Japan's population fluctuated around 31 million after 1721 and declined in the late eighteenth century. The main reason was excess mortality due to the great famines in 1732–33 and 1783–88. The famine in 1833–39 also had a significant impact on the population. These famines were caused by the cool weather, and the population decay was especially severe in the Pacific coast of northeastern Japan (Hayami 2015). Fertility decline also contributed to the stagnation and decline of the population. In the late Edo period, marriages delayed, the percentage of never-married persons increased, and marital fertility declined. Urbanization induced the demand for female labor force as housemaids, weavers, and tailors, which in turn resulted in nuptiality decline. As for marital fertility, various control methods such as abstinence, breastfeeding, abortion, infanticide, and abondance were used during the Edo period (Kito 2007).

4.7 Premodern Population Estimates of Korea

Although there is no reliable population estimate for ancient Korea, the *Book of Han* recorded that the population of Lelang commandery was 406,748 in 45 BC. If we assume that the area of the commandery was equivalent to that of Peyongan province in the Japanese era, it accounted for about 20% of the peninsula. If the population density of other areas was same as Lelang, the total population was 5 * 406,748 = 2,033,740, which works as the upper limit. As for the lower limit, the assumption that the population density of other areas was only 10% of Lelang, the total population was 1.4 * 406,748 = 569,447. It is likely that the Korean population decayed due to epidemics, wars, and famines as in ancient China. Even if the population of ancient Korea once exceeded two million, it is possible that the population collapsed to be less than a half of the peak.

As the Chinese population increased during the Tang dynasty (618–907), it is assumed that the Korean population also increased in the unified Silla period (668–935). Although there remains no population register of Silla, a famous Buddhist monk Iryeon (1206–1289) wrote in *Samguk Yusa* that Gyeongju, the capital of Silla, had 178,936 households in its golden age. This implies that the city had one million inhabitants if the average household size was larger than 5.6. While Cumings (2005) did not deny the possibility, it seems impossible that the agricultural productivity in ancient Korea could support a mega city with one million population. According to Farris (2006), the population of Kyoto in the late Heian period was about a hundred thousand, or 1.5% of the total population. If this ratio is applied to a capital with one million population, the total population of Silla was more than sixty million, which was almost equivalent to that of the Tang dynasty. This is impossible and the description of Gyeongju by Iryeon should be considered as an exaggeration. The domain of unified Silla was limited to the south of the Taedong River and was less than a half of Japan excluding Hokkaido. Even if the population density of Silla was higher than Japan in the late Heian period, the total population should have been less than five million.

Table 4.5 summarizes premodern population estimates in the Goryo and Joseon dynasties. Kim (2010b) started from the record in the *History of Song* that the adult population of Goryo was 2.1 million in 1330. She assumed that the child population was 30% of the adult population and calculated that the total population was 2.1 * 1.3 = 2.73 million. She also assumed that the adult population increased with the annual rate of 0.335% for 101 years until 1231 to be 2.94 million when the Mongolian invasion began. This growth rate seems to be valid if we consider the agricultural development in the twelfth century. According to Lee (2004), the annual cropping spread, the Champa variant was introduced, and farmlands were expanded during this century. Although Kim Soon-Ja did not write explicitly, the total population in 1231 should have been 2.94 * 1.3 = 3.822 million. Since she assumed that the adult population decayed by 46% until the second expedition to Japan in 1281, the total population in that year was 3.822 * 0.54 = 2.064 million if the rate was same for the child population. Of the total decrease, 25% was caused by the annexation of Pyongan and Hamgyong provinces to the Yuan dynasty. She found that these two provinces accounted for one quarter of the total population in 1432 and applied this proportion to the population in 1231. The remaining 21% was the net population decline due to excess mortality. The *History of Goryo* recorded that 206,800 Koreans, 7% of the total population in 1231, were captivated by the Mongol army in 1254. She simply presumed that the percentage of war deaths and that of deaths from disease during the war were also 7% without a rational basis. Furthermore, she did not consider the long-term impact of the cruel labor exploitation on the agricultural production. Thus, the net population decline could have been larger than 21%.

Yang Seong-Ji, a Korean scholar official, wrote in 1461 that the adult population of Korea was four million. Kim Soon-Ja accepted this and estimated the total population in that year was between 5.33 and 5.70 million. On the other hand, Kwon and Sin (1977) compared the population registers in 1440 with provincial reports on famine damages and concluded that the total population was 6.724 million. For the sixteenth century, they applied the coverage rate of 35.8% to the register records and estimated that the total population in 1519 was 10.469 million. They estimated that the population in 1461 was 7.564 million, which was considerably higher than

Table 4.5 Premodern population of Korea (thousand)

Year	Kim (2010b)	Kwon and Sin (1977)	Michell (1979/80)	Lee (2004)	McEvedy and Jones (1978)	Maddison (2001)
1130	2,730					
1231	3,822					
1281	2,064					
1440		6,724				
1461	5,330–5,700	7,564				
1500		9,412		9,000		8,000
1519		10,469	8,000			
1543		11,633	8,800			
1600		11,722		11,000	5,000	10,000
1650		10,889			5,000	
1681		14,651	10,900			
1693		16,030	10,200			
1700		14,359		13,500	6,250	12,200
1732		17,698	13,600			
1750		18,657	14,000		7,000	
1800		18,443		16,500	7,500	
1820		16,217	12,200			13,800
1850		16,476			9,000	
1870		16,835				14,347
1910		17,427				

that of Kim Soon-Ja. Because there remains no population register in 1543–1638, Kwon and Sin assumed that the annual growth rate was 0.4% in 1543–90 and – 0.2% in 1599–1635 without an empirical basis. Furthermore, they did not show a numerical basis for the population decline by 16.5% in 1591–98 due to the Japanese invasion. The Joseon dynasty conducted population surveys every third year after 1639. Kwon and Sin postulated that the number of households in the register was correct and that omissions occurred only for household members. They calculated the adjustment factor c with $\ln(c) = 1.8038 - 0.2105$ h, where h is the observed average household size. This implies that the correct average household size was 1.8038/0.2105 = 8.57, and a smaller h requires a large adjustment factor. Since h was around 4.2 throughout the eighteenth and nineteenth centuries, c was mostly close to $\exp(1.8038 - 0.2105 * 4.2) = 2.5$. In fact, most estimates for the late Joseon period were about 2.5 times of the registered population (Park and Cha 2004). In any case, it is difficult to accept their assumptions that the registered number of households was correct and that the average household size was constant at 8.57 for more than two centuries.

All other studies gave smaller estimates than Kwon and Sin. Michell (1979/80) also relied on the population registers but assumed higher coverage rates. According to him, the total population once exceeded ten million but decreased owing to the invasion of Japan (1592–98) and the Qing (1627, 1636–37). He assumed that the average household size was 5.3 and that the coverage rate was two thirds after the seventeenth century. He estimated the total population with the number of households * 5.3 * 1.5. His estimation suggests that the population regained ten million in the late seventeenth century but decayed again due to great famines. While the maximum population by Kwon and Sin was 18.66 million in 1750, Michell discussed that the Joseon population peaked at 14.2 million in the 1810s. Both studies suggested that the population declined by 13% in 1750–1820, reflecting the trend in the population register. Remaining three studies did not refer to the population register and showed no population decline in any time. Lee (2004) simply assumed the constant annual growth rate of 0.2% in 1500–1800. McEvedy and Jones (1978) gave smaller estimates than other studies. Especially, their estimates for 1750 and 1800 were even smaller than the population register. Maddison assumed that the population did not decline but the annual growth rate dropped from 0.22% in 1500–1600 to 0.20% in 1600–1700, 0.10% in 1700–1820, and 0.08% in 1820–70.

The fifteenth century was the golden age of the Joseon dynasty, and the population grew rapidly owing to the cultivation of coastal areas, the spread of annual cropping, the development of local medicine, and the governmental efforts for a stable agricultural production and a prevention of famine (Lee 2004). The population growth was disrupted by the Japanese and Manchurian invasions after 1592. However, neither Japan nor the Qing burned farm fields or brutally mobilized labor forces. Thus, the population decline after the end of the sixteenth century should have been considerably smaller than that in the thirteenth century. Although Michell discussed that the population declined by 20%, the magnitude was significantly smaller than that presumed by Kim Soon-Ja. While Michell asserted that it was at the end of the seventeenth century that the population regained the level before the Japanese invasion, Lee Hunchang pointed out that a loyal record in 1646 wrote that the cultivation recovered the former level. The population increase in the seventeenth century was promoted by using seedbed of rice that enabled multi-cropping. However, there is a controversy about when and why this new method spread in the southern part of Korea. Kim Jae-Ho wrote that the use of seedbed spread to cope with the cool weather in the seventeenth and eighteenth centuries. While Kuwano (2002) discussed that the seedbed was introduced as a response to the labor shortage due to the wars against Japan and the Qing, Woo (2016) asserted that the method spread after the great famines in the late seventeenth century.

Historians and demographers agree that the population growth continued in the eighteenth century, but more slowly than before. According to Lee Hunchang, the introduction of maize, potato, and sweet potato in addition to the use of seedbed of rice contributed to the population growth. Many Korean scholars agree that the population decayed in the nineteenth century as suggested in the household registers. Lee Hunchang asserted that the population decreased owing to chronic famines, frequent floods, and successive rebellions in the nineteenth century. While Park and Cha (2004) concluded from an analysis of genealogical records that the population decreased in the nineteenth century, Cha Myung-Soo later changed his mind and discussed that such facts as the declining agricultural land per capita, the falling price of slave, and the forest devastation suggest that the population growth continued into the nineteenth century (Cha 2009). In that case, the estimates without a population decline by McEvedy and Jones (1978) and by Maddison (2001) are preferable.

4.8 New Population Estimation for Premodern Korea

This section attempts a new population estimation during the Joseon dynasty because the existing estimates are too large. Table 4.6 compares the population density in 1600 suggested by the studies in Table 4.5 with regions of Japan implied in Kito (2007). The estimated population densities other than McEvedy and Jones are close to that of Northern Kyushu and are significantly higher than Japan as a whole. However, it is difficult to believe that the agricultural productivity of Korea was significantly higher than Japan in the early seventh century. Instead, it is more realistic to assume that the population density of Northern Kyushu was equivalent with southern part of peninsula, and that the northern part had lower population density. In fact, while the southern province of Jolla showed a population density of 154 persons per km^2 in the 1925 census, figures for northern provinces were 61 in Pyongan and 37 in Hamgyong. Such a regional difference possibly lasted throughout the Joseon period.

It is assumed here that the population density of Gyeongsang province was equivalent to that of Northern Kyushu in 1600. For Seoul and other provinces, the population density was obtained being based on the ratio to Gyeongsang province. Table 4.7 shows land areas taken from the 1935 census and registered populations by province

Table 4.6 Population density of Korea and Japan in 1600

Korea (person/km^2)		Japan (person/km^2)	
Kwon and Sin (1977)	53.1	Total	32.5
Lee (2004)	49.8	Kinai	223.9
Maddison (2001)	45.3	Southern Kanto	97.6
McEvedy and Jones (1978)	22.7	Tokai	48.8
		Northern Kyushu	48.4
		Hokuriku	41.6
		San'in	40.3
		Northern Kanto	37.9
		Sanyo	37.7
		Shikoku	33.3
		Tozan	24.7
		Southern Kyushu	20.3
		Western Ou	18.5
		Eastern Ou	16.1

Regional population densities of Japan are based on Kito (2007)

collected by Zensho (1927) from historical records. Population densities were calculated from these data and ratios to Gyeongsang were shown in the first panel of Table 4.8. The high ratio of Seoul reflects the high coverage of registration in the capital and the actual ratio should have been less than 10.0. The ratio in the 1925 census was 10.8 but this was a result of the rapid population growth of Seoul in the early twentieth century. The ratio of 5.0 was chosen here to adjust to the finding that the population of Seoul in the nineteenth century was about 300,000 (Kye and Park 2016).

For provinces other than Seoul, population densities relative to Gyeongsang are acceptable. Southern provinces of Jeolla and Chungcheong had densities close to Gyeongsang. The population density of Gyeonggi, the peripheral area of Seoul, increased and overtook Gyeongsang in the nineteenth century. Northern provinces of Hwanghae, Gangwon, Pyongan, and Hamgyong showed lower population densities than Gyeongsang but the difference decreased gradually. These changes are consistent with the spread of advanced agricultural methods from the south to the north. It was assumed here that the population densities of Jeolla and Chungcheong were equivalent to Gyeongsang throughout the estimation period. Other ratios shown in the second panel of Table 4.8 were adjusted so that the estimation for 1850 produced a population distribution that was close to the 1925 census.

This section presents low and high variants of estimation, which correspond to the lower and upper limits of Japan's population in 1600. Kito's regional population estimates used in Table 4.6 are consistent with Hayami's result that the total population was 12.273 million in 1600 (ISE 1974). The low variant in this section accepted Kito's estimation and assumed that the population density of Northern Kyushu was

Table 4.7 Registered population by province in Korea

Province	Land area	Registered population			
	(km^2)	1648	1753	1807	1852
Seoul	235	95,569	174,203	204,886	204,053
Gyeongsang	31,293	424,572	1,662,253	1,607,044	1,535,810
Jeolla	15,525	174,052	931,170	892,747	880,549
Chungcheon	22,441	431,837	1,192,950	1,251,069	1,068,171
Gyeonggi	12,580	81,244	642,012	674,627	672,603
Hwanghae	16,738	54,927	540,349	582,930	672,603
Gangwon	26,263	54,003	390,640	336,122	324,488
Pyongan	43,370	145,813	1,267,709	1,305,969	868,906
Hamgyong	52,325	69,348	497,446	706,012	691,655
Total	220,769	1,531,365	7,298,732	7,561,406	6,918,838

Sources Land area: 1935 census; Registered population: Zensho (1927)

Table 4.8 Ratio of population density to Gyeongsang province

Province	Observed ratio to Gyeongsang province				Assigned ratio for estimation			
	1648	1753	1807	1852	1600	1750	1800	1850
Seoul	30.03	13.98	17.01	17.72	5.0	5.0	5.0	5.0
Gyeongsang	1.00	1.00	1.00	1.00	1.0	1.0	1.0	1.0
Jeolla	0.83	1.13	1.12	1.16	1.0	1.0	1.0	1.0
Chungcheon	1.42	1.00	1.09	0.97	1.0	1.0	1.0	1.0
Gyeonggi	0.48	0.96	1.04	1.09	0.5	1.0	1.1	1.1
Hwanghae	0.24	0.61	0.68	0.82	0.3	0.6	0.7	0.7
Gangwon	0.15	0.28	0.25	0.25	0.1	0.3	0.3	0.3
Pyongan	0.25	0.55	0.59	0.41	0.2	0.3	0.5	0.5
Hamgyong	0.10	0.18	0.26	0.27	0.1	0.2	0.3	0.3

48.4 persons per km^2 in 1600, as shown in Table 4.9. The low variant assumed that the population of Gyeongsang province doubled in 1600–1800 from the same level of Northern Kyushu while Japan achieved the tripling of population. The population growth in the Joseon dynasty was supposed to have been slower than that in Tokugawa Japan because the agricultural growth and the transition to a peasant society were slower than in Japan. In 1929, three quarters of rice paddies in Korea were not irrigated when most paddies in Japan had irrigation systems (Yi 2015). While economically dependent agricultural workers mostly disappeared by the late Edo period in Japan, a slave class called "nobi" remained Korea until the early twentieth century. The population density of Gyeongsang in 1750 was interpolated with

$(2 * 48)^{0.75}$. The density in 1850 was given with 96 * 0.95 assuming that the population decreased by 5% in the first half of the nineteenth century. As shown in Table 4.4, Japan's population declined by 5% in 1721–92, and it was presumed that this level of population decay also took place in Korea.

The primary cause that induced such population decline was Confucian fundamentalism that formed the yangban culture. The Joseon dynasty enforced the family manner and ceremonies of the neo-Confucianism established by Zhu Xi. The yangban, aristocratic class, followed Confucian rules and prepared for the national examination of officials while avoiding productive manual labors. As the population of yangban increased and its lifestyle expanded, the exploitation of farmers and laborers became more and more serious. Westerners visiting Korea witnessed the low labor motivation among ordinary Koreans (Savage-Landor 1895; Bird 1898; Allen 1908). They called yangbans predators or vampires and described how they extracted and squeezed ordinary people. Thus, it was useless and even foolish to work for more than enough to support life. Such a condition, the opposite of industrious revolution in Japan, blocked economic growth and increased vulnerability to natural disasters and social conflicts.

Because the high variant corresponded to the upper limit of Japan's population in 1600, the population density of Northern Kyushu was estimated with 48.4 * 1700/1227, and the population density of Gyeongsang was assumed to have been equivalent to this value. It was assumed in the high variant that the population increased by 1.5 times in 1600–1800 since the doubling in two centuries produced a higher estimate than Kwon and Sin. Thus, the population density in 1750 was given with $(1.5 * 67)^{0.75}$. As in the low variant, it was assumed that the population declined by 5% in 1800–50. Ratios shown in the second panel of Table 4.8 were applied to the estimated population densities of Gyeongsang to produce the density and population of other provinces. According to Table 4.5, Kwon and Sin, Lee Hunchang, and Maddison agreed that the population of Joseon dynasty increased by the annual growth rate of 0.22% in the sixteenth century. Both low and high variants accepted this assumption and the population by province in 1500 was obtained by uniformly dividing the population in 1600 with the 100th power of 1.0022.

There is a controversy on the population size when Japan annexed Korea in 1910. Existing estimates raged between 15.5 million (Ishi 1972) and 17.4 million (Kwon et al. 1975; Kwon and Sin 1977). It was assumed here that the population was 16

Table 4.9 Population density of Northern Kyushu and Gyeongsang province

Nortehrn Kyushu		Gyeongsang province		
Year	Population density	Year	Low variant	High variant
1600	48.4	1600	48	67
1750	145.1	1750	80	90
1804	148.8	1800	96	100.5
1846	154.5	1850	91.2	95.5

Population density of Northern Kyushu is based on Kito (2007)

Table 4.10 Interpolation of population distribution in 1910

Province	1850	1925	1910
	Estimation	Census	Interpolation
Seoul	0.0208	0.0180	0.0186
Gyeongsang	0.2222	0.2224	0.2223
Jeolla	0.1564	0.1818	0.1767
Chungcheon	0.1144	0.1090	0.1101
Gyeonggi	0.0849	0.0841	0.0843
Hwanghae	0.0906	0.0744	0.0777
Gangwon	0.0577	0.0689	0.0666
Pyongan	0.1511	0.1383	0.1409
Hamgyong	0.1019	0.1030	0.1028
Total	1.0000	1.0000	1.0000

million for the low variant and 17 million for the high variant. To obtain the population by province, the regional distribution in 1910 was estimated. Table 4.10 compares the regional distribution in the estimates for 1850 and that in the 1925 census. Owing to the adjustment of ratios in the second panel of Table 4.8, differences between 1850 and 1925 are small and acceptable. Because the population of Seoul is known to have decreased in the late nineteenth century (Kye and Park 2016), the declining share of Seoul is understandable. If we denote the share of i-th province in year y as $s[i, y]$, the distribution in 1910 was estimated with $s[i, 1910] = s[i, 1850] + (60/75)(s[i, 1925] - s[i, 1850])$ assuming linear changes. Estimated shares were multiplied with 16 million or 17 million to produce the population by province in 1910.

Table 4.11 shows estimated populations of the Joseon dynasty by province in 1500–1910. The growth rate of a provincial population was determined by the ratios in the second panel of Table 4.8. Because the assumed trend of equalization, a province with low population density in 1600 had a high growth rate. While population of Gyeongsang, Jeolla, and Chuncheong grew by 2.8–3.2 times in 1500–1910 in the low variant, Gyeonggi's population increased by 6.2 times and Hamgyong's population by 8.7 times. In the high variant, smaller growth rates were assumed instead of larger initial populations than the lower variant. Thus, Gyeonggi's population increased by 4.8 times and Hamgyong's population by 6.6 times while Gyeongsang's population grew by 2.1–2.5 times. All the provincial populations uniformly declined by 5% in 1800–50 because same ratios were assumed in Table 4.8. Seoul's population fluctuated around 300,000 as intended, and estimates for 1910 were consistent with 342,626 in the 1925 census.

Figure 4.1 compares the total populations in two variants with studies in the past. Although three studies agreed that the population grew with the annual rate of 0.22% in the sixteenth century, the rate would be higher if the impact of Japanese invasion (1592–98) were excluded. The estimation by Kwon and Sin that the population exceeded 14 million in 1591 seems unrealistic because this implies that the population density of Korea was almost twice as much as Japan in 1600. As in Table 4.6, the

Table 4.11 Estimated population by province

Province	Low variant (thousand)					
	1500	1600	1750	1800	1850	1910
Seoul	117	145	242	290	276	297
Gyeongsang	1,244	1,549	2,582	3,099	2,944	3,557
Jeolla	875	1,091	1,818	2,181	2,072	2,827
Chungcheon	640	798	1,329	1,595	1,515	1,761
Gyeonggi	216	269	897	1,184	1,125	1,349
Hwanghae	217	271	902	1,263	1,200	1,243
Gangwon	108	134	670	804	764	1,066
Pyongan	338	422	1,054	2,108	2,002	2,254
Hamgyong	190	237	789	1,421	1,350	1,645
Total	3,945	4,915	10,284	13,946	13,248	16,000
Province	High variant (thousand)					
	1500	1600	1750	1800	1850	1910
Seoul	163	203	272	304	289	316
Gyeongsang	1,736	2,163	2,905	3,244	3,082	3,780
Jeolla	1,222	1,522	2,045	2,283	2,169	3,004
Chungcheon	894	1,113	1,496	1,670	1,587	1,872
Gyeonggi	302	376	1,009	1,240	1,178	1,433
Hwanghae	303	378	1,015	1,322	1,256	1,320
Gangwon	150	187	754	842	800	1,133
Pyongan	472	588	1,186	2,207	2,096	2,395
Hamgyong	265	331	888	1,488	1413	1,747
Total	5,507	6,860	11,569	14,599	13,869	17,000

lower limit of the population density in Japan was 32.5 per km^2 in 1600, and the upper limit was 32.5 * (17,000/12,273) = 45 per km^2. Estimates by Kwon and Sin, Lee Hunchang, and Maddison are unreliable because they exceeded the upper limit of Japan. It is incredible that the population density of Korea was higher than Japan in 1600. While Korea experienced a rapid population growth after the golden age in the fifteenth century, Japan's population also increased constantly from the mid-medieval era (1280–1450). Because the multi-cropping spread earlier, there is no reason to conclude that Japan's agricultural productivity was lower than in Korea. While the estimation by McEvedy and Jones for the population in 1600 is close to the low variant in this section, their estimates for the later periods are unrealistically small.

Table 4.12 compares the population densities of Korea implied by the low and high variants with those of Japan in 1600–1900. Ranges in 1600 are 32.5–45.0 for Japan and 22.3–31.1 for Korea. Thus, there is a possibility that the population density of Korea was almost equivalent to Japan in 1600 if we chose the lower limit of Japan and

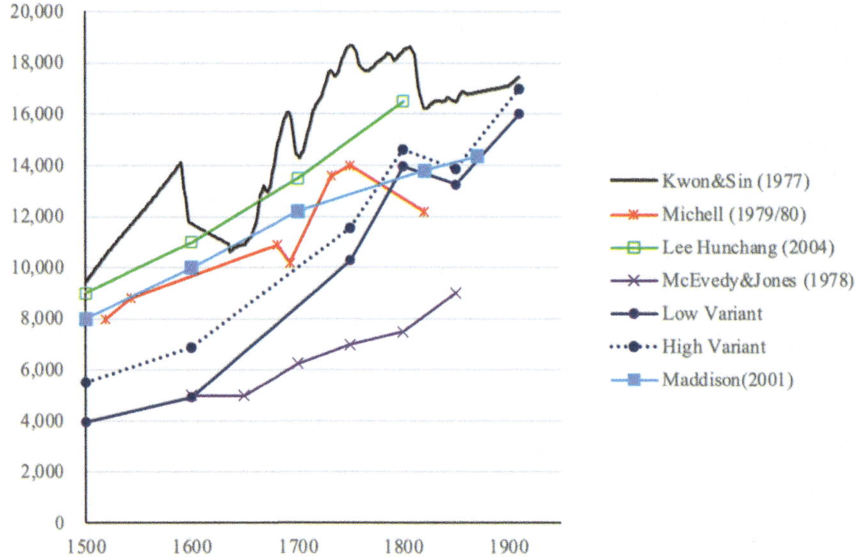

Fig. 4.1 Population estimates for Joseon Dynasty (1000)

the upper limit of Korea. However, the difference increased as Japan's population grew rapidly in the seventeenth century, and it was estimated that the population density of Korea was between 57 and 64% of Japan in 1750. Although the difference reduced as the population growth stagnated in Japan in the eighteenth century, it increased again in the nineteenth century when Japan's population began to grow again. The estimation by Kwon and Sin is unreliable because they suggested that the population density of Korea was higher than Japan until the eighteenth century. The estimates in this section for 1750 and later are close to those by Michell and Maddison. While they suggested that the population density of Korea in 1800 was about 80% of Japan, Lee Hunchang assumed that the density was 92% of Japan. However, the population density at more than 90% of Japan seems to be too high if we consider the colder weather than Japan, the undeveloped irrigation system, and the incomplete transition to peasant society in Korea.

Table 4.12 Population density of premodern Japan and Korea

Japan		Korea	
Year	(Person/km^2)	Year	(Person/km^2)
1600	32.5–45.0	1600	22.3–31.1
1750	82.0	1750	46.6–52.4
1798	80.9	1800	63.2–66.1
1846	85.5	1850	60.0–62.8
1900	123.1	1910	72.5–77.0

Population density of Korea is based on Table 4.11

The estimation in this section suggested that the average annual growth rate in 1850–1910 was 0.32% in the low variant and 0.29% in the high variant. On the other hand, the rate suggested by Kwon and Sin was as low as 0.09%. However, a relatively high growth rate around 0.3% is more realistic considering developments after the opening of Korea to foreign countries. The western medicine started to spread from Japanese settlements in Busan, Wonsan, and Incheon ports that were opened according to the Japan-Korea Treaty of 1876. After the Joseon-United States Treaty of 1882, medical missionaries were sent from western countries to Korea and accelerated the spread of western medicine and public health. In the 1890s, public health administration laws and agencies imitating those of Japan were founded. Cha (2014) confirmed that mortality was declining in the 1890s.

4.9 Premodern Population Estimates of Taiwan

Little is known about the premodern population in Taiwan. During the Dutch era (1624–61), the East India Company (VOC) invited Chinese peasants to cultivate the west coast. Koxinga's troops expelled the Dutch in 1661 and settled in Taiwan. During the Kingdom of Tungning (1661–83), the western part of the island was developed by soldiers and peasants from the mainland China. As shown in Table 4.13, it was inferred that the total population was at most 200,000 in the late seventeenth century. While the land area of Kyushu Island of Japan is approximately the same as that of Taiwan, the population of Kyushu was already exceeded 1.2 million in 1600 (Kito 2007). The low population density of Taiwan meant that there was a large room for development. Migration from the mainland China, especially from Fujian and Guangdong provinces with high population pressure, continued after the annexation to the Qing despite the government banned travelling to Taiwan. The high annual growth rate of 1.8% in 1680–1810 was achieved by those Chinese migrants.

Table 4.13 Population estimates for premodern Taiwan (thousand)

Year	Chen (1979)	Ho (1959)	Wakabayashi (2001)
1650	100		
1680	200		120–150
1811	2,000	1,902	1,940
1890	2,550	3,200[a]	
1895	2,550	2,698	

[a] 1887

4.10 Mortality Factors in Agrarian Society

The principle of population by Malthus (1798) was that a population is checked by the food production, and a famine is unavoidable if the population growth is not checked by other factors. This principle is less valid after the industrial revolution in which the food production has been growing at a higher rate than the population. In an agrarian society in which the international trade of food was small, famines were not only caused by natural disasters but also by the population pressure. Thus, it is expected that famines increased as the population approached the carrying capacity. In China, the population often collapsed at the end of a dynasty. Then, it is considered that famines were less frequent in the initial period of each dynasty when the subsistence was plentiful relative to the population. Famines increased in the end of a dynasty not only by the population pressure but also by the institutional fatigue that brought about the failure in sustaining production and distribution. Furthermore, official history books in China and Korea tend to exaggerate disasters at the end of the previous dynasty because they were written to legitimize the new dynasty.

As discussed above, famines due to the cool weather was more disastrous than that due to the warm weather, especially after the introduction of drought resistant variant. It is supposed that the damage of famine declined in a long-term because of the improvement of agricultural methods and the cumulation of wealth. While the starving people searched foods in mountains and seashores in famines until the Kamakura period in Japan, they flew into cities with plenty of wealth since the Muromachi period (Kikuchi 2000). Thus, the demographic impact due to famine was largest in the ancient era and decreased in the medieval and early modern eras. When an abnormal weather caused a poor harvest in a wide rage, there should have been a coincidence of famines in Japan, Korea, and China.

In addition to famines, Malthus considered that epidemics and wars were two major factors of positive checks that restrained the population growth by raising mortality. Especially, the first wave of a hitherto unknown disease can have a catastrophic damage as in the New Continent in the sixteenth century. McNeill (1976) inferred that the disease that killed more than 98% of population in the northwestern provinces in China in 310–12 was smallpox. Such catastrophic epidemics should have been in ancient Korea and Japan, although too old to be recorded. The plague that swept throughout Eurasia in the fourteenth century was a strain originated around Lake Issyk-Kul in Kyrgyz (Ishi 2018) and can be seen as a new disease activated by the Mongol empire. Although Syphilis killed a large number of people in Eastern Asia since the fifteenth century, a disastrous situation such as "roads were covered with corpses" was not recorded because of its chronic symptom. The pandemic of cholera in the nineteenth century was well recorded in Eastern Asia as in other parts of the world.

A lethal disease tends to be less dangerous in recurrent epidemics by the coevolution of disease orgasm and human hosts. According to McNeill (1976), smallpox and measle turned to childhood diseases by the thirteenth century. While a pandemic of plague started from Hong Kong in 1894, the death toll in 1894–1913 was only

1,245 in Japan (Iijima 2009). The dangerous classic strain of cholera was replaced by the less lethal El Tor strain in the 1960s (KSID 2009). The process of attenuation was also observed in the pandemic of COVID-19.

In an agrarian society, epidemics often prevailed during famines when the nutritional condition was worsened, and the immunity was weakened. Typhoid and typhus fever were typical diseases during the period of famine (Kikuchi 2000; Kinoshita 2002). In Japan, some famines accompanied smallpox and measle. It is more difficult to specify diseases in China because epidemics were recorded simply "yi" (epidemic) or "dayi" (great epidemic) without the name or symptom in official history books. While the appendix table of McNeill (1976) contained 290 cases of epidemics in China between 243 BC and 1911, he could not give the name of disease for most cases. The name of disease was rarely recorded also in Korea, and some traditional names are difficult to identify. Although it was considered that "jang-yeok" in the Goryeo period was malaria, KSID (2009) was skeptical about this hypothesis. KSID agreed that "on-yeok" in the Goryeo period was typhus and "ak-byeong" prevailed in Hwanghae province in the Joseon period was epidemic encephalomyelitis.

Malthus chose a war as a major mortality factor considering the situation of Europe where wars were frequent. In Eastern Asia, however, wars between nations were less frequent, especially after the Ming dynasty. The Ming suffered several invasions from the north in its early period and fought with Japan in Korea at the end of the sixteenth century. The Qing dynasty conducted military campaigns in Outer Manchuria, Xinjiang, Taiwan, Viet Nam, Myanmar, and Nepal before the First Opium War. Japan did not fight with a foreign force for more than 260 years after the invasion to Korea. In China, peasant rebellions at the end of the Ming dynasty and in the late eighteenth century had larger demographic impacts than wars against foreign countries. Although the Joseon dynasty suffered invasions in the sixteenth and seventieth centuries, there was no war or civil war until a series of peasant rebellions in the nineteenth century. In Japan, there was no major civil war other than Siege of Osaka (1614–15) and Shimabara Rebellion (1637–38) throughout the Edo period, and natural disasters might have killed more people than peasant rebellions or other battles.

Wars and rebellions often accompanied famines and epidemics, intensifying the demographic impact. In China, famines induced peasant rebellions and military campaigns stimulated epidemics. Such coincidences of famines, rebellions, and epidemics caused the population collapses at the end of the Yuan and Ming dynasties. The Mongol army took the scorched earth strategy and cruelly mobilized labor forces in Korea, which caused great famines and epidemics that collapsed the population of Goryeo. In the Joseon dynasty, invasions by Japan and the Qing also accompanied famines and epidemics. In Japan, a coincidence of a civil war, famine, and epidemic can be found in the period of the Genpei War (1180–85). It is known that the number of deaths from disease often exceeded the number of deaths from the battle itself. During the Japanese invasion to Korea in 1592–98, a large number of deaths from illness occurred in the Japanese army, the Ming army, and the Korean navy.

4.10.1 Premodern Major Epidemics

Fujikawa (1912) collected more than 300 cases of epidemics between the sixth and nineteenth centuries in Japan. Especially disastrous cases were chosen and shown in Table 4.14. Although Fujikawa did not include, the *Nihon Shoki* recorded that there was a catastrophic epidemic killing more than half the population under the reign of Sujin, an emperor assumed to reign in the fourth century. It is possible that the unrecorded first wave of lethal diseases such as smallpox, measle, influenza, or plague had even more disastrous impact. Early epidemics of eruptional diseases are difficult to determine whether they were smallpox or measle. While many historians considered that epidemics in 552 and 558 were those of smallpox, some hypothesized that they were measle. It was since the eighth century that smallpox and measle were distinguished and recorded with different names. Fujikawa listed 58 cases supposed to be smallpox and 47 cases of measle.

It has been assumed that Japan was free from the plague because it is isolated from the continent and there was no large colony of mice or other rodents (Yasuda 1995; Maddison 2001). As a fact, there remains no record of prominent plague symptoms such as lymphatic malignancy or black tissues. However, McNeill (1976) asserted that the epidemic in 808 that killed over half the population could have been plague because there was an epidemic of plague in coastal region of China in 762–808. Dong (2014) agreed that the disease was more likely to have been plague than smallpox or influenza considering the warm winter in 807. While the plague prevailed in China in the mid-fourteenth and late sixteenth centuries, there is no record in Japan that suggest overseas transmission. For the epidemic of plague in the mid-seventeenth-century China, severe epidemics in 1642 were recorded in Korea and Japan, suggesting international transmissions of plague. However, there is no record of symptom to identify the kind of disease.

While Fujikawa included 41 cases of influenza epidemics, Kurachi (2016) contained five epidemics in the Edo period that were not in Fujikawa's table. Large death tolls were recorded for the epidemic in Kyoto from the winter of 862 and that in Edo in 1716. According to Fujikawa, no typhoid epidemic can be identified before the Edo period. He found twelve cases of typhoid epidemics between the sixteenth and the nineteenth centuries. The epidemic in 1674 was especially disastrous because it concurred with famine.

The first cholera pandemic started from Kolkata in 1817 and spread globally by the British ships (McNeill 1976). In Japan, cholera landed on Kyushu in autumn of 1822 and advanced eastward. There remains a controversy on the route of cholera invasion to Japan. While Fujikawa (1912) and Horiguchi (1972) asserted that a Dutch ship came from Jakarta to Nagasaki transmitted cholera, the KSID (2009) concluded that the disease was transmitted from Busan to Nagasaki via Tsushima. Although the first wave did not reach Edo, the second wave in 1858–59 caused a tremendous number of deaths in Kanto area including Edo (Ishi 2018). The death toll in Edo recorded in literatures reviewed by Fujikawa ranged between 28,000 and 268,000.

Table 4.14 Major epidemics in premodern Japan

Year	Disease	Description
4C?	?	More than half the population died with epidemic
552	Measle? small pox?	Nationwide epidemic
585	Small pox?	Deaths from eruption throughout the nation
706	?	A large number of deaths due to epidemic
735	Small pox	"Beans eruption" broke out from summer to winter
737	Small pox	Uncountable deaths due to eruption disease
790	Small pox	A large number of deaths due to "beans eruption"
808	Plague?	A large number of deaths due to famine and disease
862–63	Influenza	A large number of deaths due to "cough disease" from winter
929–30	?	Epidemic in Kinki area, roads were covered with corpses
994	?	More than half the population died with nationwide epidemic
998	Measle	A large number of deaths due to "mottled eruption"
1001	Small pox	Roads were covered with uncountable corpses
1040	?	A large number of deaths in Kyoto due to long lasting epidemic
1044	?	Roads were covered with corpses due to epidemic from January to June
1135	?	Roads were covered with corpses due to epidemic and famine
1277	?	Roads were covered with corpses due to epidemic since autumn
1360	?	Roads were covered with corpses due to epidemic and famine
1421	?	Uncountable deaths due to epidemic and famine
1424	?	Families and villages died out due to epidemic and famine
1450	?	More than 1000 died per day in Kyoto due to epidemic from January to July
1460–61	?	Roads were covered with corpses due to epidemic and famine
1486–87	?	More than half the population died with epidemic
1492	?	Roads and fields were covered with corpses due to epidemic and famine
1511	Tonsillitis	Uncountable deaths due to "mouth paralysis"
1540	?	A large number of deaths due to nationwide epidemic
1642	Plague?	A large number of deaths due to nationwide epidemic
1674	Typhoid	Uncountable number of deaths due to "wet disease" in Kinki
1684	?	Thousands of deaths due to epidemic started in Nagasaki
1716	Influenza	More than 80,000 people died in Kanto
1732	?	More than 70,000 people died in Kyushu
1773	?	More than 190,000 people died in Edo between March and May
1822	Cholera	More than 3000 people died in Hagi city due to strange disease since August

(continued)

Table 4.14 (continued)

Year	Disease	Description
1858–59	Cholera	More than 40,000 people died in Edo between July and September
1862	Measle and cholera	Measle in June and July, "violent diarrhea" in August

Selected from Fujikawa (1912)

An even larger death toll occurred in 1862 when smallpox in spring was immediately tailed by cholera in summer.

While the KSID (2009) included more than 800 cases of epidemics between 15 BC and 1944, records before the Goryo dynasty were too brief to know the extent of damages. Epidemics with a large number of deaths in the Goryo and Joseon dynasties were chosen and shown in Table 4.15. As in China, official history books in Korea did not record disease names but simply wrote "yeok" ("yi" in Chines) or "yeo" ("li" in Chinese). Because both words mean an epidemic of acute disease, "yeoyeok" does not help to specify the disease. Furthermore, an epidemic that lasted long could have included more than one disease. For example, the epidemic since 1717 had peaks in summer of 1717, spring in 1718, and spring in 1719. This shows a possibility that a summer disease such as typhoid was followed by a spring disease such as influenza or measle.

Existing literatures in Korea did not specify disease names other than measles and cholera. The epidemic prevailed in the capital of Goryeo in 1030 was thought to have been measles transmitted from Japan (KSID 2009; Lee 2010). In 1281, soldiers and sailors came back from the second Mongol expedition to Japan caused a disastrous epidemic of measles in northern China and Korea (Lee 2010). The chronological table in the KSID (2009) exceptionally specified that the epidemic in Hwanghae province in 1707 was measles. Cholera came from Manchuria to Pyongan province in September 1821 and killed several hundred thousand of people within a month. It is estimated that one million people died in Korea in this first prevalence of cholera (Lee 2010).

It is assumed here that a severe epidemic ravaged Korea when plague prevailed in China could have been plague. According to KSID (2009), corpses covered roads of the capital of Goryeo due to a severe pestilence in April 1348. McNeill (1976) pointed out the possibility that the epidemic that killed nine tenth of population in Hebei province in 1331 was plague. Cao (1997) discussed that the disease that collapsed the population of Shanxi province in 1353 was also plague. In fact, the plague rampaged in various parts of Eurasia for several decades in the early and middle fourteenth century, as in the case of the Black Death in Europe (1346–53). Thus, it is at least possible that the disaster in the Kaesong city in 1348 was the plague transmitted from China.

According to Cao (1997), the plague that broke out in Shanxi province in the 1560s forwarded east and prevailed in Hebei and Shandong provinces in the 1580s. In

Table 4.15 Major epidemics in premodern Korea

Year	Disease	Description
1030	Measle?	A large number of deaths in December in Kaesong
1109–10	?	A large number of death due to nationwide epidemic
1162	?	Land was covered with corpses due to famine and epidemic in May
1254–56	?	Roads were covered with corpses due to epidemic
1281	Measle?	A large number of deaths due to epidemic from spring to autumn
1348	Plague?	Corpses heaped in Kaesong due to epidemic in April
1383	?	Famine and epidemic in March, navy soldiers mostly died out
1524–26	?	A large number of deaths due to nationwide epidemic of "yeoyeok"
1554	?	A large number of deaths due to nationwide epidemic
1577	?	A large number of deaths due to nationwide epidemic
1587–88	Plague?	Thousands of deaths in Hwanghae in May
1612	?	A large number of deaths due to nationwide epidemic
1627	?	A large number of deaths due to great famine, war, and epidemic
1643–44	Plague?	A large number of deaths due to nationwide epidemic
1653–55	?	A large number of deaths due to nationwide epidemic
1670	?	A large number of deaths due to great famine and epidemic
1688	?	More than 10,000 deaths due to nationwide epidemic
1697–99	?	A large number of deaths due to great famine and epidemic
1707	Measle	More than 10,000 deaths due to measle in Hwanghae in April
1714	?	Thousands of deaths in Jeju
1717–19	?	A large number of deaths due to nationwide epidemic of "yeoyeok"
1732–33	?	A large number of deaths due to nationwide epidemic of "yeoyeok"
1741–43	?	A large number of deaths due to nationwide epidemic of "yeoyeok"
1750	?	A large number of deaths due to nationwide epidemic of "yeoyeok"
1770	?	A large number of deaths due to nationwide epidemic of "yeoyeok"
1775	Measle	A large number of deaths due to measle in July
1793	?	A large number of death due to epidemic in Chungcheong and Gyeongsang
1799	?	128,000 deaths in Seoul due to epidemic in January
1821–22	Cholera	A large number of deaths in Pyongyang and Seoul due to strange disease

Selected from KSID (2009), and Lee (2010)

Korea, thousands of people died due to an epidemic in Hwanghae province in 1587–88 (KSID 2009). Cao (1997) wrote that plague rampaged in a wide range of Hebei including Beijing in 1640–41. Among cases of epidemics included in the chronological table of KSID (2009), pestilences in Seoul (December 1642 and November 1643), in Jeolla province (December 1643), and the nationwide pandemic in February 1644 have the possibility of plague. As discussed above, a disastrous epidemic took place also in Japan in 1642.

While the appendix table of McNeill (1976) contained more than two hundred cases of epidemics in China between 243 BC and 1911, most descriptions were too brief to know the disease name or the extent of damage. Disastrous epidemics with some details were chosen and shown in Table 4.16. McNeill suggested that epidemics in 161–12, in 311–12, and in 322 represent the first arrival in China of some hitherto unknown infection such as smallpox or measles, although he avoided a clear identification.

McNeill confirmed that epidemics with high mortality rates in 762 and 806 were plagues. It is assumed here that catastrophic pestilences in 1331–58 were plagues. Cao (1997) checked clinical descriptions in local history records and identified that the epidemic in 1641 was plague. An outbreak of cholera in the hinterland of Calcutta in 1817 was transported by British ships to Sri Lanka, Southeastern Asia, and Eastern Asia (McNeill 1976). Cholera landed on Guangdong province in 1820 and spread to Fujian and Zhejiang provinces. In 1821, the pestilence leaped from Nanjing to Beijing, and eventually covered whole China (KSID 2009). The epidemic in northern Taiwan in 1820 could be the cholera transmitted from mainland China (Liu and Liu 2001).

4.10.2 International Transmission of Infectious Disease

Table 4.17 summarizes international transmissions of infectious diseases in Eastern Asia. As already mentioned, the catastrophic pestilence in the fourth-century Japan probably represents the arrival of new disease. History books in Japan recorded that smallpox came from Baekje together with the Buddhism in 552. KSID (2009) also argued that smallpox was transmitted from Shilla to Japan in 735–37. The transmissions of plagues and cholera were already discussed above.

During the Goryeo period, two new diseases called "jang-yeok" and "on-yeok" came from the Song dynasty. While KSID (2009) was skeptical about the hypothesis that "jang-yeok" ("zhang-yi" in Chinese) was malaria, no alternative hypothesis was given. While the name "on-yeok" ("wen-yi" in Chinese) is used as synonym of epidemic in China today, the word was used to refer to typhus in Japan. Although both diseases are unidentifiable, they did not give serious damages as measles or plague. Scarlet fever was transmitted from the Ming to the Joseon dynasty in 1613. At that time, a famous physician named Heo Jun (1539–1615) identified the disease and established a treatment method (KSID 2009).

According to an essay cited in Fujikawa (1912), an unidentified disease broke out in Nagasaki in April 1684 and the death toll rose over one thousand. In June, the disease ravaged in Kyoto and Osaka, killing several thousand people. Since Nagasaki was the only harbor opened to China and the Netherland, it is reasonable to guess that the disease was transmitted by a foreign ship. By the seventeenth century, the cyclical prevalence of measles came to be known in Japan and Korea. A comparison of Fujikawa (1912) and KSID (2009) showed that measles prevailed in Japan and Korea with one-year delay in 1691–92, 1729–30, 1752–53, and 1802–03,

Table 4.16 Major epidemics in premodern China

Year	Disease	Description
161–12	?	Epidemic broke out in ranks of army in Xinjiang and Qinghai; three or four out of ten died
310–12	Smallpox?	Epidemic, locality undefined, following on earlier disasters from locusts and famine, northern and central China became a “great wasteland”; in Shaanxi only one or two out of a hundred taxpayers survived
322	?	Epidemic; two or three out of ten died; location undefined
423	?	Epidemic in North China, in Henan two or three out of ten died
468	?	Epidemic throughout the empire; during a second outbreak later in the year in Henan, Hebei, Shandong, Hubei, and Anhui 140,000–150,000 died
682	?	Epidemic in Henan and Shandong; land covered with corpses
762	Plague	Epidemic in Shandong; more than half the population died
806	Plague	Epidemic in Zhejiang; more than half the population died
891	?	Epidemic in Hubei, Jiangsu and Anhui, in Hubei three or four out of ten died
1127	?	Epidemic in Henan; half population of capital died
1232	?	Epidemic in Henan; 90,000 died in less than fifty days
1275	?	Epidemic with incalculable mortality, location undefined
1308	?	Epidemic in Zhejiang; more than 26,000 died
1331	Plague	Epidemic in Hebei; nine tenths died
1351–52	Plague	Epidemic in Shanxi, Hebei, Jiangxi, 50% mortality among troops in the Huai Valley
1353	Plague	Epidemic in Hubei, Jiangxi, Shanxi, Suiyuan; in part of Shanxi more than two thirds of the population died
1354	Plague	Epidemic in Shanxi, Hubei, Hebei, Jiangxi, Hunan, Guangdong, and Guangxi. In part of Hubei six or seven out often of the popu lation died
1358	Plague	Epidemic in Shanxi and Hebei, over 200,000 died
1369	?	Epidemic in Fujian; corpses in heaps on the roads
1408	?	Epidemic in Jiangxi, Sichuan, and Fujian; 78,400 died
1489	?	Epidemic in Hunan; whole villages and towns perished
1562	?	Epidemic in Fujian; seven tenths died
1618	?	Epidemic in Shanxi, Hunan, Guizhou, and Yunnan; corpses lying side by side in Shanxi
1641	Plague	Epidemic in Henan, Hebei, Shandong, and Shanxi; corpses lying side by side throughout
1757	?	Epidemic in Zhejiang and Shanxi, in Xijiang, on the western border, everyone afflicted with the disease died without exception
1819–22	Cholera	Cholera landed on Guangdong and spread nationwide

Selected from McNeill (1976), place names were changed to pinyin

Table 4.17 International transmission of infectious diseases in premodern Eastern Asia

Year	Disease	Transmission	Source
4C?	?	? → Japan	–
552	Smallpox	Baekje → Japan	KSID (2009)
735–37	Smallpox	Shilla → Japan	KSID (2009)
807–08	Plague?	? → Japan	McNeill (1976)
1018	Jang-yeok (malaria?)	Song → Japan	KSID (2009)
1030	Measles	Japan → Goryeo	KSID (2009)
1100	On-yeok (typhus?)	Song → Goryeo	KSID (2009)
1331	Plague	Mongol → Hebei	McNeill (1976)
1348	Plague?	Yuan → Joseon	–
1587–88	Plague?	Ming → Joseon	–
1613	Scarlet fever	Ming → Joseon	KSID (2009)
1642–44	Plague?	Ming → Joseon, Japan?	–
1684	?	? → Japan	–
1691–92	Measles	Japan → Joseon	–
1729–30	Measles	Joseon → Japan	–
1752–53	Measles	Joseon → Japan	–
1801–02	Influenza	Batavia → Japan	Fujikawa (1912)
1802–03	Measles	Joseon → Japan	Ishi (2018)
1820–22	Cholera	Qing → Joseon, Japan	KSID (2009)

as shown in Table 4.17. Ishi (2018) wrote that measles was transmitted from Korea to Japan in 1803 but he did not mention to earlier cases. An influenza epidemic started in Nagasaki in December 1801 and reached Kyoto and Osaka in February 1802. According to Fujikawa, Japanese physicians at that time assumed that the disease was brought by a Dutch vessel.

4.10.3 Broad Range Famines in Eastern Asia

In Japan, great famines are called with era names such as the "Yowa Famine." Official history books in China and Korea distinguished great famines ("da-ji" in Chinese, "dae-gi" in Korean) from ordinary famines ("ji" in Chinese, "gi" in Korean). Two catastrophic famines in Korea were named after Chinese zodiac. The Gyeong-shin Famine in 1670–71 was named because 1670 was the year of "gyeong-sul" ("geng-xu" in China, "kanoe-inu" in Japan), and 1671 was the year of "shin-hae" ("xin-hai" in China, "kanoto-i" in Japan). For the Eul-byeong Famine in 1695–99, 1695 was the year of "eul-hae" ("yi-hai" in China, "kinoto-i" in Japan), and 1696 was the year of "byeong-ja" ("bing-zi" in China, "hinoe-ne" in Japan).

Table 4.18 shows great famines with era names in Japan and those occurred in the same period in Korea and China. While there was no great famine in Japan in 1670–71 when the extremely disastrous Gyeong-shin Famine hit Korea, it is impressive that the Genroku Famine in Japan and Eul-byeong Famine in Korea took place in the same year. This implies that sometimes an abnormal weather in a broad range of areas can cause simultaneous famines across countries. This section examines if famines in different countries were caused by a common climate change.

The Yowa Famine was caused by the extremely small amount of rain in western Japan in 1180. The *History of Song* recorded great droughts in Jiangsu, Zhejiang, Jiangxi, Anhui, Hubei, and Hunan provinces in 1180, and great famines in three of them. It is possible that the amount of rain was extremely small in a broad range from Changjiang basin to Japan in this year. While the Oei Famine was caused by a drought in 1420, the cause of great famine in Shandong province in the same year cannot be specified from the *History of Ming*. The Tenmon Famine occurred because of the heavy rain, flood, and locust plague in 1539. The *Veritable Records of the Joseon Dynasty* also recorded that a great famine hit Korea due to the rain and snow in spring and drought and locust in summer of 1539. Those weather conditions possibly stimulated the outbreak of locust in Japan and Korea, resulting in great famines in both countries. The main causes of Kan'ei Famine were the drought in summer and the heavy rain in autumn of 1641. The Korean peninsula was already hit by a great famine due to the drought in 1640, and the famine continued into 1641

Table 4.18 Great famines in Japan and concurrent great famines in Korea and China

Japan	Korea	China
Yowa Famine (1180–82)		Great famine in Jiangsu, Zhejiang and Anhui (1179–81)
Kanki Famine (1229–32)		
Shoka Famine (1229–32)		
Oei Famine (1420–21)		Great famine in Shandong (1420)
Kamakura Famine (1428)		
Kansho Famine (1459–61)		
Tenmon Famine (1539–40)	Great famine (1539, 41–42)	
Kan'ei Famine (1641–43)	Great famine (1640–41)	
	Gyeong-shin Famine (1670–71)	Great famine in Zhejiang (1671–72)
Genroku Famine (1695–96)	Eul-byeong Famine (1695–99)	
Kyoho Famine (1732–33)		
Horeki Famine (1755–56)		
Tenmei Famine (1783–84)		
Tenpo Famine (1833–39)	Floods and famines (1832–39)	Great famine in Shandong (1833)

because of the drought and locust plague. Although the *History of Ming* recorded no great famine for 1641, ordinary famines caused by droughts hit Hebei, Shandong, and Jiangsu provinces.

The Gyeong-shin Famine in Korea occurred because of the drought in May, the frost and hail in June and July, and the snow in August of 1670. The locust plague was added and destroyed the harvest of crops, and the epidemic accelerated excess mortality (Woo 2016). Although damages of floods or cold summer were recorded in Japan, there was no great famine. According to the *Draft History of Qing*, a famine in 1670–71 was recorded only in Jiaxing city in Zhejiang province, while there were some damages of droughts and locust plagues in Hebei, Zhejiang, and Anhui province. Thus, it seems that the extreme weather in 1670 was limited to the Korean peninsula. On the other hand, both Korea and Japan suffered from the catastrophic harvest failure due to the cold summer in 1695. Although the *Draft History of Qing* recorded no great famine in 1695, there was an ordinary famine in Hebei province due to heavy rain. The Tenpo Famine in Japan was caused by the heavy rain and cold summer in 1833. Korea also had damages of floods and rains in 1833, and eventually suffered from a famine in the next spring. The great famine in Shandong was not the result of the abnormal weather in 1833 but caused by the frost and flood in the previous year. However, there was an ordinary famine in Hebei province due to the drought in spring and the rain and hail in autumn of 1833.

4.10.4 Epidemics and Famines During Wars and Rebellions

The population of China sometimes collapsed due to famines and epidemics concurrent to wars or rebellions. Table 4.19 lists such correspondences of Malthusian positive checks. However, causes of early population crises such as those in the Qin-Han replacement or in the late Eastern Han are difficult to determine because of the sparse historical records.

The disaster in 310–12 happened in the order of famine, war, and epidemic. The Western Jin started declining immediately after the unification of China in 280. The political confusion was accelerated by the civil war called the War of the Eight Princes (291–306). According to the *Book of Jin*, locusts ate up all the crops in a broad range of areas including Hebei, Henan, Shanxi, Shaanxi, Gansu, Jiangsu, Anhui, and Hubei in the summer of 310. In October, a serious famine hit Luoyang, the capital of the Western Jin. In April 311, the royal army had a crushing defeat against the Former Zhao and more than 100,000 soldiers died. In June, the army of the Former Zhao invaded Luoyang, victimizing a large number of citizens. The food prices escalated due to the famine in this month. In January 312, Emperor Huai of Jin escaped from the capital, and battles between the Western Jin and the Former Zhao continued in various places. According to McNeill (1976), a great famine hit many provinces in 312, and in Shaanxi only one or two out of a hundred taxpayers survived.

As shown in Table 4.2, the population declined sharply in the early seventh century when the Tang replaced the Sui. This population crisis was caused by excessive labor

Table 4.19 Epidemics and famines during wars and rebellions in China

War, rebellion	Famine	Epidemic
War between Western Jin and Formar Zhao (311–12)	Locusts and great famine in Hebei, Henan, Shanxi, Shaanxi, Gansu, Jiangsu, Anhui, and Hubei (310)	Great epidemic making northern and central China a "great wasteland," in Shaanxi only one or two out of a hundred taxpayers survived (312)
Sui-Tang replacement (612–19)	Famine, cannibalism prevailed and villages were ruined (618)	A large number of deaths due to drought and epidemic (612)
Liao's invasion to Later Jin (944–46)	Great famine (944, 946)	
Jin's invasion to Northern Song (1126–27)	Great famine in Kaifeng, skeletal corpses covered roads (1127)	Great epidemic, half of Kaifeng citizens died (1127)
Yuan's invasion to Jin (1231–33)	More than one million died due to famine and epidemic (1232)	
Red Turban Rebellions (1351–66)		Great epidemic in Hebei, Shanxi, and Jiangxi (1352)
		Great epidemic in Hubei, Jiangxi, Shanxi, and Suiyuan (1353)
	Great famine in Hebei, Jiangxi, Fujian, and Huguang (1354)	Great epidemic in Hebei, Shanxi, Jiangxi, and Huguang (1354)
		Great epidemic in Henan (1356)
	Great famine in Henan (1357)	Great epidemic in Shandong (1357)
	Great famine in Hebei and Shandong (1358)	Great epidemic in Hebei and Shanxi (1358)
	Great famine in Hebei, Henan, and Shandong (1359)	Great epidemic in Shaanxi, Shandong, and Guangdong (1359)
		Great epidemic in Jiangsu, Zhejiang, and Anhui (1360)
	Great famine in Hebei (1361)	Great epidemic in Zhejiang (1362)
Peasants' rebellion in late Ming dynasty (1628–45)	Starving peasants revolted in Shaanxi (1628)	
	Great famine in Shanxi (1634–35)	Plague in Shanxi (1633–44)
	Great famine in Henan, Shandong, and Shanxi (1636–37)	Plague in Shaanxi (1636–40)

(continued)

Table 4.19 (continued)

War, rebellion	Famine	Epidemic
	Famine in northern and central China (1640)	Plague in Hebei, Henan and Shandong (1640–44)
	Famine in Hebei, Shandong, Jiangsu, and Zhejiang (1641)	

mobilization for the construction of the Grand Canal and the expedition to Goguryeo, which resulted in a series of revolts. According to the *Book of Sui*, a large number of people died for famine and epidemic in 612, the year when the campaign against Goguryeo was conducted. In 618, while the civil war was at its peak, a famine hit the country and many villages ruined. Although the population recovered during the Tang dynasty, it decayed again in the late Tang and the Five Dynasties and Ten Kingdoms periods. In addition to battles between the five dynasties, northern China suffered invasions by the Liao dynasty. For example, the Liao attacked the Later Jin in northern China in 944 and 946. According to the *New History of Five Dynasties*, there were great famines in both years. On the other hand, the appendix table of McNeill (1976) contains no epidemic record in the Five Dynasties and Ten Kingdoms period.

Kaifeng in Henan replaced ruined Xi'an in Shaanxi to become the political center in the late Tang period. Four of the Five Dynasties including the Later Jin chose Kaifeng as the capital. Although the Liao's invasion in 944 and 946 did not reach Kaifeng, there should have some damages from great famines. Kaifeng also served as the capital of the Song dynasty until the Jin demolished the Northern Song in 1127. Although the Jin dynasty chose Beijing as the capital and transferred from Harbin in 1153, the capital relocated again to Kaifeng in 1215 under the pressure of Mongol invasion. Kaifeng was the site of horrible sieges at the end of two dynasties and a large number of people died for famines and epidemics. The *History of Song* recorded that when the Jin occupied Kaifeng in January 1127, a half population of the capital died for epidemic, and corpses on roads had almost no fresh due to famine. According to the *History of Jin*, Henan province suffered famine in 1232, and hunger spread in the capital when the Mongol army surrounded in March. In June, cold weather aroused epidemic and more than 90,000 died in less than fifty days.

The Red Turban Rebellions broke out in Henan in 1351 and eventually led to the collapse of the Yuan dynasty in 1368. The military power of Mongol decayed rapidly due to the prevalence of plague since 1331, and the imperial army had no power to suppress the rebellions. While Table 4.16 listed catastrophic epidemics with some detailed records, the *History of Yuan* simply wrote that there were great epidemics in Hebei in 1356, in Shandong in 1357, in Shaanxi, Shandong, and Guangdong in 1359, in Jiangsu, Zhejiang, and Anhui in 1360, and in Zhejiang in 1362. The great famine in 1354 harmed a broad range from Hebei to Huguang (Hubei, Hunan, Guangdong, and Guangxi), and cannibalism was reported in various places. Great famines occurred successively since 1357 in northern China where fierce battles between the imperial and rebel armies continued. The *History of Yuan* wrote that the death toll of famine in Hebei was uncountable in 1359, corpses of starved people covered roads, and soldiers captivated weakened persons to eat.

The peasants' rebellion in the late Ming period was induced by the droughts in Shaanxi province in 1627–28. While the rule of Ming ended in April 1644 with the occupation of Beijing by the rebel army led by Li Zicheng, the Qing army drove out him in October and started an occupation policy of China. In Shanxi province, there was a great famine due to drought and an epidemic of plague in 1634–35. Cao (1997) assumed that poor harvest with drought stimulated migration of rodents and encouraged the spread of plague. The drought in Shanxi continued into 1636–37, and there was a great famine due to locusts in Henan and Shandong. According to Cao, the plague in Shanxi was immediately transmitted to Shaanxi province. In 1640, famines due to droughts and locusts swept a broad range including Hebei, Henan, Shandong, Shanxi, Shaanxi, Jiangsu, and Zhejiang. Famines continued into the next year in Hebei, Shandong, Jiangsu, and Zhejiang. The rebel army of Li Zicheng invaded Henan in 1640 and ravened about the province in 1641–42. However, Li was pushed back to the south by the Qing army that invaded Hebei and Shandong in 1642. In 1643, Li's army invaded Shaanxi and occupied Xi'an in October. In 1644, Beijing was successively occupied by Li and Qing. According to Cao, the pestilence in Hebei, Henan, and Shandong were stimulated by those military campaigns, and presumably it was Li's army that transmitted the pathogen from Shanxi to Hebei.

As for the Mongol invasion of Korea, the damage was especially large in the 1250s when famines and epidemics were added to the war. According to the *History of Goryeo*, in January 1251 Gojong (reign 1213–59) grieved the devastation of northern region saying that it looked like a no man's land. In 1253, the Mongol army stormed through Goryeo and sent a diplomatic document recommending a surrender writing "there were thousands of deaths every day due to our military campaign." There was a drought in April 1254 and an epidemic in the capital in June in addition to brutal battles. The *History of Goryeo* wrote "the Mongol army captivated more than 206,800 males and females, the death toll was uncountable, and cities were completely destructed." While the war continued from the beginning of 1255, the nation fell into a great famine in the spring, and a great epidemic broke out again in the capital in December. Battles went on in 1256, and corpses covered roads due to a great famine and epidemic in the winter. Although the war ended in 1259, the capital was struck by a great famine and epidemic, and refugees heading the south distracted roads and many high-class officials died. While Kim (2010b) considered only war deaths and that from disease during the war, total excess mortality including epidemics and famines could have halved the population of Goryeo.

Famines and epidemics were recorded also during Japan's invasion of Korea in the late sixteenth century. The Japanese army landed on Busan in April 1592, occupied Pyongyang in June, and stopped there. The Ming army recaptured Pyongyang in January 1593, and the Japanese army retreated to Seoul. According to the Imperial Japanese Army General Staff Office (1924), there were many deaths from starvation in February 1593 due to famine, especially in Gyeongsang, Gyeonggi, and Hwanghae provinces where the Japanese army passed in the previous year. Possible causes may include the evacuation of peasants abandoning agricultural works, and the disobedience of slaves due to the collapsed social order. Epidemics prevailed since March was possibly caused by the migration of armies. Because the Japanese army had difficulty to continue battles due to the hunger and disease, they retreated from Seoul to the southern tips of peninsula in April. Although the Ming army attempted to pursuit the Japanese in May, they could not continue to fight either. Famines and epidemics were not limited to the armies but seriously harmed ordinary Koreans. The famine continued to 1594 and cases of cannibalism were recorded in various places. Korea could have a good harvest in 1595, but a famine revisited in 1596 in addition to the prevalence of smallpox possibly carried by the Ming army. The war was restarted in 1597, and there were battles in Gyeongsang, Jeolla, and Chungcheong provinces. The war ended in 1598 when Toyotomi Hideyoshi died in August.

The Yowa Famine occurred in the early stage of the Genpei War fought between the Genji (Minamoto clan) and the Heike (Taira clan) in 1180–85. The famine was caused by the drought in the spring and summer of 1180, and hunger and disease spread nationwide in 1181. The poor harvest due to the drought continued in this year, and streets of Kyoto were covered with corpses in January 1182. On the other hand, the Genpei War was caused by the falling power of Heike, and troops of Genji rose in 1180 in eastern and central Japan marched to Kyoto. The Genji army occupied Kyoto in 1183, and the Heike fled westward. After several battles to secure the command of the Seto Inland Sea, the war ended with the victory of the Genji in 1185. Because battles in this period was moderate and small, the war might not add a significant death toll to the Yowa Famine.

While the Onin War (1167–77) was historically important because it ruined Kyoto and opened the door to the warring states period, its demographic impact was possibly small because it was not accompanied by a great famine or epidemic. Although there was a prevalence of measles in 1471 and that of smallpox in 1477, descriptions cited in Fujikawa (1912) are not serious. Unlike in China and Korea, it is difficult to find a war that caused a great famine, or a famine that resulted in a war in Japan.

4.10.5 Demographic Impacts

As can be seen in the estimation of Goryeo population by Kim (2010b), it is difficult to evaluate an impact of population disaster. As for the case of plague in the late sixteenth-century China, Cao (1997) applied a description in *Xinle Xianzhi* that 40% of inhabitants died to his estimated population of Hebei and Shanxi in 1582 and obtained the death toll of $17.8 * 0.4 = 7$ million. He asserted that even this was an overestimation, at least 5 million (28%) died. Since there was no major famine or battle in this period, the gross decrease of population must have been close to this net impact of plague. However, the decline by 28% still seems to be an overestimation. It is assumed here that the plague in the fourteenth-century China was the first wave after the Mongol army unfolded the pathogen in central Asia and that the population declined by one third as in Europe. The plague in the sixteenth century was the second wave, and that in the seventeenth century was the third wave. While the fourth wave started from Hong Kong at the end of the nineteenth century and caused a global pandemic, the death toll accounted for only 1% or less in each country. If we consider that the impact of plague reduced from 33.3% in the first wave to 1% in the fourth wave, a linear assumption gives the population decline by 22.6% and a log-linear assumption gives that by 10.4% in the second wave. Iijima (2009) also judged that Cao's result was an overestimation, although he did not give any alternatives.

In the Yi Joseon dynasty, the population register was updated every third year. Appendix Table 1 of Kwon and Sin (1977) included the total registered population in 92 years between 1393 and 1861. This implies that only 92/468/3 = 6.6% of registration records is available today. The data for the early Joseon period are especially rare and there remain only eight population registers before 1638. Periods of 1441–1519 and 1544–1638 are totally vacant. Thus, there is no numerical basis to evaluate the demographic impact of the Japanese invasion in 1592–98. For the period of the Gyeong-shin Famine (1670–71) and that of the Eul-byeong Famine (1695–99), the population declines can be evaluated with the registered population. There was a significant decrease in the registered population in 1811–16, which was supposed to be caused by Hong Gyeong-Rae's rebellion (1811–12) and famines. According to Kwon and Sin, there were floods in Hamgyong and Gyeongsang, and famines in Gyeonggi, Chungcheong, Gyeongsang, and Gangwon provinces in 1814.

Table 4.20 summarizes population declines in two major famines in the seventeenth century and in the riot and famine in the nineteenth century. The total registered population decreased by 8.6% in the Gyeong-shin Famine, by 20.1% in the Eul-byeong Famine, and by 13.0% in the 1810s. Estimates by Kwon and Sin (1977) showed smaller change than the population register, which could be a result of an inadequate smoothing. The Gyeong-shin Famine and the Eul-byeong Famine were so disastrous that the Veritable Records of the Joseon Dynasty recurrently referred to as "dae-gi" (great famine). The population decrease by 1.3% is too small for such a catastrophic disaster. Although the registered population decayed by 8.6%, the number of households decreased by 10.1%. Thus, there remains a possibility that the population declined by 10% or more by the Gyeong-shin Famine. Woo (2016) also estimated that 10% of the total population died for hunger and diseases. For the Eul-byeong Famine, the change in the register and the estimation by Kwon and Sin can be seen as the upper and lower limits. Woo discussed that the death toll of the Eul-byeong Famine was about one million, which was equivalent with the Gyeong-shin Famine. If the total population in 1695 was greater than that in 1670, as suggested in the estimates of Kwon and Shin, the rate of excess death should have been less than 10%. It is difficult to believe the result of Kwon and Sin that Hong Gyeong-Rae's rebellion and famines in 1812–14 caused a more serious damage than two great famines. If the recorded population decline was an exaggeration due to an administrative problem, the actual decrease rate should have been less than 5%.

Kwon and Sin (1977) uniformly applied the annual growth rate of – 2.58% for 1592–98 and suggested that the population decreased by 16.5% from 14.1 to 11.8 million by the Japanese invasion. On the other hand, Woo (2016) asserted that the Gyeong-shin Famine and the Eul-byeong Famine had larger demographic impacts

Table 4.20 Demographic impacts of great famines in premodern Korea

Famine	Year	Population register			Estimation by Kwon and Sin (1977)		
		Population	Change	(%)	Population	Change	(%)
Gyeong-shin Famine	1670	5,165	– 444	(– 8.6)	13,106	– 171	(– 1.3)
(1670–71)	1672	4,721			12,935		
Eul-byeong Famine	1693	7,045	– 1418	(– 20.1)	16,030	– 831	(– 5.2)
(1695–99)	1696	5,627			15,199		
Rebellion and Famine	1811	7,583	– 988	(– 13.0)	18,305	– 1577	(– 8.6)
(1812–14)	1816	6,595			16,728		

than the Japanese invasion. In such a case, the decrease of population was less than 10%.

In Japan, while the population register system called "shumon aratame-cho" was established in 1671, the total populations are available only after 1721 when the first registration census was conducted. The census was conducted every sixth year after 1726, and the national populations are known for fifteen years shown in Table 4.4. Thus, it is impossible to compare the demographic impacts of the Genroku Famine and the Eul-byeong Famine happened in the same period.

Table 4.21 shows the demographic impact of great famines that occurred in the eighteenth and nineteenth-century Japan. Although the decrease in the registered population was always less than 5%, actual impact should have been larger if the minimum population after the famine were available. However, it is unlikely that the population declined by 10% or more as in the seventeenth-century Korea. It is considered that damages of famines decreased in a long-term in Japan owing to the economic development and relief policies. Although there is no numerical evidence, damages of famines in the seventeenth century also should have been mild due a rapid agricultural growth. Thus, it is thought that a great famine as catastrophic as that in Korea was possible only in the ancient and medieval Japan.

Japan consists of volcano islands and is often hit by earthquakes and tsunamis. Those temporal and local natural disasters did not result in a significant decline in the national population unless they accompanied famines or epidemics. However, it is likely that more people died for natural disasters than for civil wars or peasants' rebellions during the peaceful Edo period. The table of historic tsunamis by Tsuji (2006) contains four great tsunamis of 30 m high in 869, 1611, 1707, and 1771. The Jogan earthquake in 869 occurred in the bottom of the Pacific Ocean at the east of Tohoku region, and a great tsunami hit a broad range of eastern Japan. Table 4.22 summarizes disaster death tolls described by Kurachi (2016). The Sanriku earthquake in October 1611 accompanied a great tsunami that attacked Tohoku region. It was recorded that 1783 persons died in Sendai Domain, and that more than 3000

Table 4.21 Demographic impacts of great famines in premodern Japan

Famine	Year	Population register		
		Population	Change	(%)
Kyoho Famine	1732	32,306	– 1205	(– 3.7)
(1732–33)	1750	31,101		
Horeki Famine	1756	31,285	– 179	(– 0.6)
(1755–56)	1762	31,106		
Tenmei Famine	1780	31,213	– 1343	(– 4.3)
(1783–84)	1782	29,870		
Tenpo Famine	1834	32,477	– 1375	(– 4.2)
(1833–39)	1840	31,102		

people and horses died in today's Iwate and Aomori prefectures. The epicenter of the Genroku earthquake (1703) was the sea bottom near Nojimazaki in Chiba prefecture and caused great damages to the Kanto and Tokai areas. Odawara Domain in today's Kanagawa prefecture suffered especially serious damage due to the fire in the cattle town in addition to the quake and tsunami.

It is estimated that the magnitude of the Hoei earthquake in October 1707 was M8.6, which was the strongest in the historical period of Japan. The damages of quake and tsunami struck broad areas between Kanto and Kyushu. Although the death toll reported to the shogunate summed up to 5045, some feudal lords did not report the number of deaths. Thus, the actual figure should have been several times larger than the reported death toll. The Hoei eruption of Mt. Fuji, which was possibly the most explosive eruption in historical Japan, occurred 49 days after the earthquake. It was fortunate that the eruption happened during winter and did not lead to a great famine. On the other hand, the eruption of Mt. Asama in July 1783 caused a severe damage to the harvest and worsened the damage of Tenmei Famine (1783–84). The eruption of Mt. Unzen in April 1792 resulted in a landslide and tsunami killing 15,135 people in Nagasaki and Kumamoto prefectures.

Table 4.22 Death tolls from disasters in Edo period Japan

Year	Disaster	Death toll
1611	Sanriku Earthquake	1783 in Sendai Domain
1657	Great Fire of Meireki	106,046 burn deaths
1703	Genroku Earthquake	10,367
1707	Hoei Earthquake	5045
1792	Unzen Landslide and Tsunami	15,135
1847	Zenkoji Earthquake	8000
1855	Ansei Edo Earthquake	10,000

Selected from Kurachi (2016)

The Zenkoji earthquake in March 1847 directly hit today's Nagano city and the death toll was elevated by the flood of the Chikuma River in addition to the quake and fire. While a series of major earthquakes struck Japan during the Ansei era (1854–60), the earthquake in October 1855 with the magnitude of M7.0–7.1 directly struck Edo city. Approximately 10,000 people died due to house collapses and subsequent fires. While a direct hit of earthquake on a large city caused a horrible demographic impact, a great fire independent of a natural disaster also recorded an awful death toll. Damages of fires were especially serious in Edo where wooden houses were densely built, and air is dry during winter months. The death toll of 106,046 from fire was recorded in the great fire in January 1657, and more people died from cold in the heavy snow after the fire. Edo suffered great fires also in 1772 and 1806, each killing several thousand inhabitants.

4.11 Premodern Demographic Indicators

Premodern demographic rates of birth, death, marriage and divorce are more difficult to obtain than the population size or distribution. In Japan, reliable vital rates are available only after the establishment of population register system called "shumon aratame" in the seventeenth century. The survey of shumon aratame began in today's Nagasaki prefecture to prohibit Christianity. The shogunate ordered feudal lords to conduct the religious survey in 1644 and enforced annual survey in 1671. All the households other than those of samurai class were annually surveyed and changes in membership due to birth, death, marriage, divorce, leaving, and returning were recorded. The survey results of each village were compiled into a shumon aratame-cho and submitted to the lord or governor. This makes possible to easily obtain the population by sex and age group in a village. Thus, a shumon aratame-cho is more advantageous for a historical demographic study than a parish register in Europe because one can calculate vital rates and migration rates from one source (Kito 2000). The shumon aratame records often used by demographers include those of Nishijo in Gifu (Tomobe 2001; Ochiai 2009; Kurosu et al. 2012), Yokouchi in Nagano (Hayami and Uchida 1972), Kabutoyama in Saitama (Kito 2000), Niita and Shimomoriya in Fukushima (Hirai 2008; Ochiai 2009), and Nomo in Nagasaki (Tsuya 2001a; Kurosu et al. 2012). In addition to a shumon aratame-cho, a "kainin kakiage-cho" (pregnancy record) is used for fertility and a "kako-cho" (family register of deaths) is used for mortality studies.

While the population and household register in the Joseon dynasty was made for a long period between the fourteenth and nineteenth centuries, its accuracy has been questioned because of the low coverage and apparent misreports. Even with correct information, it would be difficult to estimate demographic rates from waves of survey with three-year interval. However, Shikata (1976) asserted that the registration records are still usable if a researcher has sufficient knowledge on Korean history. In addition to the registration of Daegu that Shikata analyzed, those of Danseong in Gyeongsang province and of Seoul have been also used by Korean demographers

(Kim 2016a, b; Kye and Park 2016). Some historians prefer genealogical records to population registrations even though the former is limited to upper class males that survived until marriage. Park and Cha (2004) asserted that it is easier to adjust the bias in a genealogy book than in the case of population register. Both registers and genealogies are also used by Chinese historians. Lee and Campbell (1997) analyzed the *Household and Population Registers of the Eight Banner Han Chinese Army* in Liaoning province. The genealogy record of the Qing imperial lineage is famous for its good quality (Caldwell 2001).

Table 4.23 compares demographic indicators estimated in those historical demographic studies in Japan, Korea, China, and Western Europe. A unique marriage pattern characterized with late marriage and high proportion remaining unmarried prevailed in Western Europe between the eighteenth and the early twentieth centuries (Hajnal 1965). It is supposed that women married earlier, and spinsters were less common in non-European societies in the premodern period. In fact, Eastern Asian societies showed early and universal marriage patterns, although some region in Japan had a pattern close to Europe. Japan's similarity to Europe and the difference from Confucian societies were apparent in mortality and fertility. The contrast between Western Europe and Japan on one side and Confucian societies on the other is most obvious in life expectancies estimated by Maddison (2001). Such a demographic contrast can be added to the political, economic, and cultural differences between Japan and Confucian countries.

Harari (2011) discussed that the reason why Japan successfully followed up Europe in the late nineteenth century was that the Japanese leaders in the Meiji period paid efforts to westernize various parts of society. However, more important reason was that Japan was already resembled to Europe before the Western impact. Common features include the historical development from feudalism to absolutism (Eisenstadt 1996), the leadership of warrior class (Cumings 2005), the emergence of bourgeoisie (Umezao 1967), the appearance of the decentralized political system (Wittfogel 1959; Hayami 2015), and high trust on non-family members (Fukuyama 1995). How the premodern Japanese family differed from the Confucian family was discussed in Goode (1963) and Suzuki (2013, 2019).

According to Hajnal (1965), the European marriage pattern characterized by the late marriage and the high proportion remaining single existed at least for two centuries up to 1940 in the west of a line drawn between Saint Petersburg (Leningrad when he wrote it) and Trieste. He discussed that in the European pattern, the female mean age at first marriage was above 23, and was in general above 24. Although he did not refer to the lower limit of men, the youngest mean age at first marriage in his literature survey was 24.1 for men and 23.1 for women. Thus, 24 and 23 were shown in Table 4.23 as lower limit of men and women, respectively. Mortality declined in eighteenth century Europe owing to the growth in agricultural production and the immune acquisition against major infectious diseases. It is possible that the European marriage pattern emerged as a response to the increasing population pressure, and Malthus (1798) approved the delayed marriage as the only acceptable way to curb the population growth.

Table 4.23 Demographic indicators in 18th and 19th century

	Western Europe	Japan	Korea	China
Age at first marriage (male)	> 24[a]	18–29[b]	16–18[b]	21–22[b]
Age at first marriage (female)	> 23[a]	14–23[b]	16–19[b]	17–19[b]
Crude birth rate (‰)	28–38[c]	18–40[d]	30–45[e]	
Total marital fertility rate	7.5–9.0[f]	3.7–7.5[g]	6.5–9.2[h]	6.0–7.5[i]
Total fertility rate	4–6[i]	2.8–5.8[j]	5.3–8.0[h]	5–6[i]
Crude death rate (‰)	20–36[c]	16–37[d]	30–35[e]	
Infant mortality rate (‰)	200[k]	200[k]		
Life expectancy at birth (both sexes)	36[l]	34[l]	23[h]	21[l]
Percentage of urban population	20–50[m]	12–14[g]	4?[n]	< 12[o]

[a]Hajnal (1965)
[b]Kim (2016b)
[c]Glass (1965)
[d]Tsuya (2002)
[e]Kwon and Kim (2002)
[f]Lee and Wang (1999)
[g]Kito (2007)
[h]Cha (2009)
[i]Willson (2001)
[j]Kito (2001)
[k]Kito (2000)
[l]Maddison (2001)
[m]Davenport (2020)
[n]Kim (1992)
[o]Okamoto (2015)

Marriages in Eastern Asia in the eighteenth and nineteenth centuries stayed early and universal, as in other non-European areas. The range of the mean age at marriage reviewed by Kim (2016b) was wider in Japan than in Confucian countries. Marriage was especially early in the Tohoku region of Japan, and the mean ages at marriage in Niita and Shimomoriya in the early eighteenth century were 20.8 and 13.7 for male and female, respectively (Hirai 2008). On the other hand, the mean ages in western Kyushu were 31.1 and 25.3 for male and female, respectively, which were as late as in Europe (Fauve-Chamoux and Ochiai 2009). The difference resulted from the custom of servant works. While men and women worked as servants after their marriage in Tohoku, unmarried people left home to work in western Kyushu (Kurosu et al. 2012). No area where marriage of a woman was as late as in Europe was found in China or Korea. Unlike in Japan, premarital intercourse was strictly prohibited in Confucian societies. Thus, parents forced children to marry as early as possible to prevent the dishonor of the family (Caldwell 2001).

In the European marriage pattern, the age difference between husband and wife was small and couples in which the wife was older than her husband accounted for a considerable part. In Japan, the age difference was larger and there were less wives older than husbands (Kurosu et al. 2012). According to Kim (2016a), men in Danseong area in the eighteenth century married as early as women and the age difference was even smaller than in Europe. In the urban area of Daegu, the average age at marriages of male and female were exactly same, and the percentage of wives older than husbands was higher in the upper class (Kim 2016b). A wife older than husband was also preferred in wealthy families in China. While parents of wealthy families wanted to marry their sons as early as possible, they did not want a daughter-in-law who was too young to support their family members (Wolf and Huang 1980).

Divorce rates in the eighteenth and nineteenth-century Japan were among the highest found in premodern societies. In contrast with Catholic societies in Europe and Confucian societies in Eastern Asia where divorce was rare, Japan and some Southeastern Asian countries had high divorce rates (Fauve-Chamoux and Ochiai 2009). According to Campbell and Kurosu (2018), regions with high divorce rates included Taiwan, Malaysia, and Java. The life course analysis by Kurosu (2009) revealed that more than a half of first marriages ended in divorces in Shimomoriya and Niita. Unlike in China and Korea, Japanese women remarried immediately after the divorce or the death of husband because marriage was considered as the most important social security. The analysis of same villages by Hirai (2008) showed that 11.35% of all marriages were terminated by divorce. While a daughter-in-law resided with parents of her husband, her membership was ambiguous for several years. By the early nineteenth century, the full membership of a daughter-in-law was approved after the fifth year of marriage, and a divorce was avoided.

England succeeded in the simultaneous and continuous growth of production and population owing to the industrial revolution since the mid-eighteenth century. This modern population growth was achieved by rising fertility rather than declining mortality (Livi-Bacci 2012). As a result, the Crude Birth Rates (CBRs) of England and Wales in 1710–1870 reviewed in Glass (1965) were, on average, higher than those in premodern Japan. The vital rates collected by Tsuya (2002) ranged between today's Iwate prefecture and Nagasaki prefecture. Premodern CBRs tended to be low in Northeastern Japan and high in Southwestern Japan. Kwon and Kim (2002) discussed that the CBR was within the range of 30–45‰ and the Crude Death Rate (CDR) was 30–35‰ during the Joseon dynasty. Although no reliable estimation of vital rate was found for historical China, the situation was same as in Korea that the high birth rate was required to compensate the high death rate.

The European way of curbing fertility with late marriage was not found in premodern societies outside of Europe. In non-European societies, not nuptiality but marital fertility was restricted through abstinence including non-coital sexual activity, abortion, infanticide, and neglect of young children. As a result, the Total Marial Fertility Rates (TMFRs) of Japan and China were apparently lower than Europe. While TMFRs estimated by Kito (2007) ranged between 3.69 and 7.48 (Table 4.23), the national average for Japan was 5.69. Although the estimates for Nobi (today's Aichi, Mie, and Gifu), Hokuriku (Fukui, Ishikawa, Toyama, and Niigata) and Kinai exceeded 7.0, they were still lower than low fertility countries in Western Europe. The estimated TMFR of 3.69 for Tohoku region was apparently at the below-replacement level. The estimation by Hirai (2008) for Niita was 2.77, and the village population decreased from 540 to 366 in 1720–1820. The combination of high nuptiality and low marital fertility in Japan resulted in TFRs within the range of Western Europe. The national average of TFRs estimated by Kito (2001) was 4.53, which was lower than the Western European average. China also showed early and universal marriage of women and low marital fertility. As Wilson (2001) noted, the TMFR was possibly between 6.0 and 7.5 for most historical Chinese populations, which was at the low end of the European range of experience. Such low marital fertility was combined with high nuptiality and produced moderate overall fertility, with TFRs between five and six.

Cha (2009) estimated from his analysis of genealogical records that the TMFR of Korea increased from 6.45 in the eighteenth century to 9.20 in the nineteenth century, and that the TFR rose from 5.27 to 7.98 in the same period. His estimation of fertility in the nineteenth-century Korea was higher than other demographers and consistent with his conclusion that the population continued to increase. His estimation of the CBR in the nineteenth century was 62.9‰, which was out of the range assumed by Kwon and Kim (2002). Because the estimation of Cha is based on genealogical records limited to yangban adult males, it is more difficult to accept than outcomes for Japan and Europe based on materials of better quality. It is more likely that the population stagnated, and fertility was at the similar level of China in the nineteenth-century Korea.

Marital fertility in historical China was significantly lower than in Europe due to infanticide and neglect. Because more girls were victimized than boys, males tended to marry later and remained unmarried (Livi-Bacci 2012). Lee and Campbell (1997) estimated that about 30% of baby girls born in Daoyi village died from infanticide or neglect by the age of five years. As a result, 29% of males in late twenties remained single, contrasted with 2.5% of females in the same age group. Although infanticide and neglect were also conducted in Japan, premodern farmers wanted to have children of both sexes. In Shimomoriya and Niita, while the percentage of never-married women in late twenties was 2.1% which was about the same as Daoyi, the percentage for men was 10.5% which was significantly lower than China (Tsuya 2001b). Park (2016) asserted that infanticide was rare in premodern Korea because there remains practically no historical record referring to such an activity. However, it is unlikely that the Korean population sustained both high nuptiality as in Japan and high marital fertility as in Europe. Although the detail is unknown, premodern Koreans limited their marital fertility in some way.

As shown in Table 4.23, the CDRs of Japan in eighteenth and nineteenth centuries were approximately at the same level as those in England. There was only one case among CDRs collected by Tsuya (2002) that exceeded 30‰, which was observed in today's Iwate prefecture. While Kwon and Kim (2002) assumed that the CDR of the Joseon dynasty was between 30 and 35‰, it seems to be too low. In a stationary population, the reciprocal of the life expectancy at birth is equivalent with the CDR. If the life expectancy was between 20 and 25 years in the Joseon dynasty, the CDR should have been between 40 and 50‰. It is likely that China's CDR was also at this level.

The life expectancy at birth in 1820 estimated by Maddison (2001) was 36 years for Western Europe, 34 years for Japan, and 21 years for China. Estimates for Tokugawa Japan by Japanese demographers reviewed by Kito (2007) were slightly longer than Maddison, mostly raged in 35–39 years. The life tables made by Hirai (2008) from the pooled data of Shimomoriya and Niita implied even higher life expectancy at birth; 42.5 years and 41.9 years for male and female, respectively. The life expectancy for premodern Korea in Table 4.23 is the estimation by Cha (2009) based on genealogical records. According to Wilson (2001), the life expectancy at birth in historical China was generally estimated as being in the high twenties, which is again longer than Maddison. The life tables of Daoyi created by Lee and Campbell (1997) showed that the life expectancies were 35.9 years and 29.0 years for male and female, respectively. Thus, it is possible that mortality of China was lower than Korea even though it was very likely higher than Japan.

According to Kito (2000), the infant mortality rate (IMR) of premodern Japan was about the same level as England and lower than the European average. The IMR of 200‰ means that one fifth of newborn babies could not survive until the first birthday. Although there is no reliable estimate for Korea and China, the IMR was certainly higher than 300‰ in the nineteenth century. Such high infant mortality was observed even in the twentieth century. According to the life tables for Korea in 1906–10 estimated by Ishi (1972), the probability of dying until the first birthday (${}_1q_0$) was 0.335 and 0.277 for male and female, respectively. The life tables for rural China around 1930 made by Barclay et al. (1976) showed that the probability was 0.288 and 0.305 for male and female, respectively. Thus, it is likely that the IMR of Confucian countries was higher than 300‰ in the nineteenth century, and even higher in a disaster year or in an area where infanticide was common.

In Western Europe, urbanization moderately proceeded during the recovery from the Black Death. In England and Wales, the percentage living in towns with a population of 5000 or more grew from less than 10% in 1600 to more than 20% in the early eighteenth century. Urbanization was accelerated after the industrial revolution, and over half the population lived in settlements of 2500 or more by 1851 (Davenport 2020). In Japan, about 5% of the population lived in what might have been called cities in 1600 (Farris 2006). While the percentage rose to 17% in 1650, it declined to 13.6% in 1750 and 12.2% in 1850 (Kito 2007). This happened due to the population decline in middle and large cities other than Edo, which was caused by the protoindustrialization that motivated capital investment and cumulation in rural areas instead of existing cities. On the other hand, the population of Edo increased dramatically in the seventeenth century. The samurai population concentrated in Edo owing to the alternate attendance system called "sankin kotai." The population of Edo exceeded one million in the mid-Edo period to be the world's most populated city at that time (Hayami 2015).

Kim (1992) estimated the populations of 57 cities of Korea based on the household register in 1879. According to his results, the total urban population was 782,794, which accounted for 4.35% of the national population estimated to be 17,977,098. However, his result did not follow the rank-size rule and the population of Pyongyang, the second largest city, was only 14% of Seoul. It seems that he could not compensate the low coverage rates of cities other than Seoul. If C_i is the population of i-th largest city, and if the population of cities follow the rank-size rule with the intercept of log C_1 and slope of -1, then $C_i = C_1/i$ holds. If we accept that the population of Seoul was 233,702 as estimated by Kim, the total population of 57 cities should have been 1,081,810. Since Kim's national population seems too large, it is assumed here that the total population of late eighteenth-century Korea was about 13.5 million from the results shown in Table 4.11. Then, the percentage of urban population should be 8%, which is higher than Kim's result but considerably lower than Japan. While 4.35% seems too low even under the undeveloped economy in the Joseon dynasty, 8% would be more realistic. The percentage of urban population in the nineteenth-century China could have been higher than Korea but possibly lower than Japan. According to Okamoto (2015), the number of towns with a population of 5000 or more was smaller relative to Japan and most Chinese lived in communities with a population less than 5000.

Premodern cities were characterized by high mortality due to the dense dwelling situation and low fertility due to the high percentage of unmarried people. Most cities had negative rate of natural increase and had to sustain the population by absorbing in-migrants from surrounding rural areas (Kito 2007). Hayami (2015) asserted that the population growth rate in economically developed areas such as Southern Kanto and Kinai was low in the eighteenth and nineteenth century because of the urbanization. According to Ito (1995), the negative natural growth in cities of Japan lasted until the end of the nineteenth century. It is likely that the "urban graveyard" situation continued longer in other parts of Eastern Asia than in Japan.

4.12 Premodern Household

In the nineteenth-century Japan, the modal household type was a stem family in which the eldest son remained in the parental home after marriage and succeeded the home business. Because the succession of business and asset was more important than the continuity of patrilineal line, an unrelated son-in-law was adopted as a designated heir and succeeded the family asset when there was no adequate son. The impartial inheritance by a son or a son-in-law and the stem family household emerged in the upper class in the late twelfth century (Banse 2015). Such rules of inheritance and household formation prevailed among farmers' families in Kinai by 1500 and in central and western Japan by the mid-sixteenth century (Farris 2006). In Tohoku region, however, the Japanese family "ie" characterized with the desire of perpetuation, the impartial inheritance, and the stem family household emerged as late as in the early nineteenth century (Hirai 2008).

By the medieval period, the size of rural household varied in a wide range and some households contained more than ten persons including domestic servants and farm workers. After the transition to a peasant society, most agricultural production was executed by small family units and the mean household size converged to around five persons. Hayami and Uchida (1972) showed the detailed data of population and

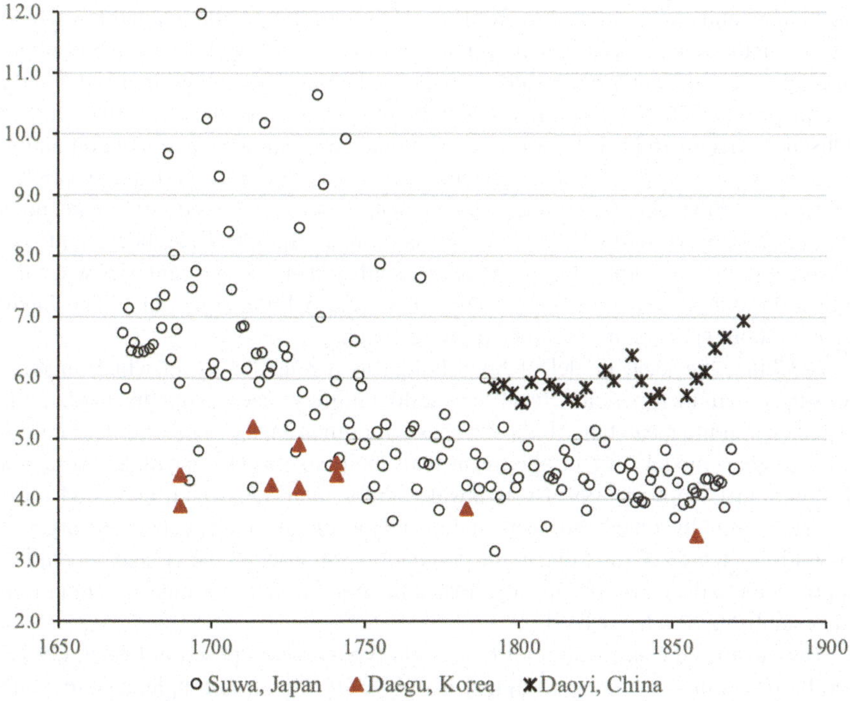

Fig. 4.2 Mean household size in premodern Eastern Asia

households in 1671–1870 for thirty-eight villages in Suwa county, Nagano prefecture. The county averages of household size were calculated and shown in Fig. 4.2 with the data in Korea and China. While the mean household size over six persons was often observed until the early eighteenth century, the mean size converged to around five after the mid-eighteenth century.

According to Nakane (1972), the mean household size in preindustrial Japan was five or a little under in most rural villages. Because of the one-son succession rule, a household rarely included lateral kins. Under high premodern mortality, a multiple family household including two-generation couples could not last long. Exceptionally large mean sizes were observed in Takayama town and Shirakawa village in today's Gifu prefecture. In those communities, a newly married couple not in the line of succession could not establish a new household due to the restricted residential area or the necessity of the family business of sericulture. While a wealthy mercantile house in Edo or Osaka included many employees as household members, such live-in servants mostly disappeared in other cities by the mid-nineteenth century (Hamano 2007).

The household register of Daegu city in Gyeongsang province has been studied well and the mean household size was presented in Shikata (1976), Shima (2004), and Kim (2016b). According to their results, the mean was less than five persons in most years and was smaller than in Suwa. However, the population and household register in the Joseon dynasty is notorious for low coverage and the mean household size can be significantly undervalued owing to the omission of children and elderly members who were not taxed. Kwon and Sin (1977) assumed that the actual average household size in 1670–1861 was 8.57 when they estimated the national population from the population register. However, that value seems to be too high as the national average. In Suwa county, the mean larger than 8.5 was observed only eight times among 139 observations. Furthermore, it is unlikely that the mean household size stayed at such a high level during the seventeenth and eighteenth centuries when the transition to a peasant society proceeded in Korea. A little larger value than Japan such as about six persons would be more realistic.

In China, the ideal model of household was a joint family in which married sons lived with their parents. Because wealthy families materialized this model, it is expected that the mean household size was larger than Japan where the stem family was the ideal model. In fact, the mean household size in Daoyi village presented by Lee and Campbell (1997) was always larger than 5.5, which exceeded most observations in the nineteenth-century Japan. Households in Daoyi became larger in the mid-nineteenth century due to the population pressure. As the village population approached to the carrying capacity, it became more difficult for men to get married and establish a new household.

The most used classification of household types was proposed in Laslett (1972) and based on the concept of Conjugal Family Unit (CFU). A couple, a couple and offspring, a parent and offspring can form a CFU. A simple family household contains CFU members only. An extended family household consists of one CFU and other kin members. A multiple family household includes two or more CFUs. In premodern Japan, the household composition differed considerably by region, and the proportion of multiple family household was low in Nishijo village in central Japan and high in Shimomoriya village in Tohoku region. Although not shown in Table 4.24, the periodical change was also significant. According to Hirai (2008), the percentage of multiple family household in Niita village increased from about 30% in the eighteenth century to 60% in the 1860s. However, lateral kins were rarely included in extended and multiple family households due to the social norm of inheritance and coresidence.

Table 4.24 Distribution by household type (%)

	Nishijo	Shimomoriya	Jinju	Daoyi
	Japan	Japan	Korea	China
	1800	1800	1831	1801
Simple family	41.3	22.2	42.8	23.3
Extended family	36.3	19.4	28.8	14.2
Multiple family	11.3	43.1	20.4	48.6
Other	11.1	15.3	8.0	13.9

Ochiai (2009), Inoue (2005), Lee and Campbell (1997)

According to Inoue (2005), simple family households dominated in Socheonri, Jinju city in Gyeongsang province. However, a modal household could have been more complex considering household members omitted from the population register. In Daoyi village, there were many multiple family households extended horizontally owing to the ideal model of joint family. A horizontal household composed of a head and his married brothers tended to divide soon after the death of the head. However, after 1810, it became difficult for married brothers to separate and establish a new household. Thus, "diagonal" households that included CFUs of uncles and nephews increased (Lee and Campbell 1997).

According to Shikata (1976), the number of slaves per household in Daegu increased from 0.54 persons in 1690 to 1.37 persons in 1858. While those slaves called "nobi" lived independently by the eighteenth century, they were impoverished and absorbed into yangban households in the nineteenth century. Economically dependent nonrelative members were rare in households in China and Japan where the transition to a peasant society was more thoroughgoing than in Korea. According to Lee and Campbell (1997), tenancy and hired farm hands were largely uncommon in China. Instead, most households absorbed distant relatives as laborers. Laslett (1972) showed that the mean number of servants in Nishinomiya, western Japan, was 0.2 and smaller than western societies. In Niita village, the mean number of servants fluctuated around 0.5 through the eighteenth and nineteenth centuries (Hirai 2008). According to Wall (1995), while the number of servants in 18 villages of England was 0.51 in 1750–1821, it was as high as 2.13 in London in 1695. Although the mean number of servants also should have been high in commercial areas of Edo and Osaka, such areas were exceptional.

4.13 Summary and Conclusion

As discussed in Chap. 3, Eastern Asian populations increased by 40–90 times during the agrarian period. However, the population growth often stagnated for centuries and sometimes collapsed. In China, peasant rebellions at the end of a dynasty often accompanied famines and epidemics and resulted in a disastrous population decline. In the thirteenth-century Korea, the Mongol invasion and the cruel labor exploitation caused a catastrophic population crisis. No population collapse comparable to those in China or Korea was found in Japan.

A population growth accelerated when a society transitioned to an annual cropping or a multi-cropping, when farmlands were expanded, and when a new crop or variant was introduced. The Chinese population grew rapidly in the twelfth and sixteenth century owing to the introduction of the Champa rice and American food plants, respectively. The boom of river construction in the sixteenth and seventeenth centuries caused a substantial population growth in Japan. A new population estimation for premodern Korea was given in this chapter because existing estimates were supposed to be overestimated. The basic assumption was that the population density of Korean provinces other than Gyeongsang, Jeolla, and Chungcheong in 1600 was lower than that of northern Kyushu.

It was inferred that damages of famine and epidemic decreased in a long run. Although the demographic impact of epidemics was largest at the first visit, the earliest transmission of the smallpox or measles from China to Korea or Japan was too old to be recorded. While those eruptional diseases were recurrently transmitted from China, there were cases that the measles was transmitted from Japan to Korea. Although there is no record of plague symptoms in Japan or Korea, some disastrous epidemics were possibly the plague transmitted from China. The cholera was transmitted from China to Korea in the 1820s. The route to Japan is still controversial.

A simple explanation of why Japan could achieve the modernization and quickly caught up the Western powers is that Japan was already resembled to Europe in the eighteenth century. While the political power structure and the importance of nonfamily organizations were pointed out, demographic similarities such as low fertility, low mortality, and advanced urbanization can be added to the list. Mortality in the eighteenth and nineteenth-century Japan was as low as that in England, and fertility was lower because the modern population growth already started in England. China and Korea needed high fertility to compensate higher mortality than Japan or Europe. Although the percentage of urban population declined in the eighteenth-century Japan, it was still higher than that in China or Korea.

The mean household size declined to five or less while Suwa county in Japan transitioned to a peasant society in the eighteenth century. Although the household registration of Daegu in Korea showed small household sizes, it was inferred that the mean household size was actually larger than Japan. Households of Daoyi in China was larger than Japan or Korea because of the inclusion of lateral kins. Owing to the economic difficulty in the late nineteenth century, households became even larger and more complicated. It is likely that households in the nineteenth-century Korea contained more servants than those in Japan or Europe. Thus, simple and small households without a large number of lateral kins or unrelated servants can be added to the list of similarity between Japan and Western Europe.

References

Allen, Horace G. 1908. *Things Korean: A Collection of Sketches and Anecdotes Missionary and Diplomatic*. Fleming H. Revell Co.

Banse, Tomomi. 2015. Nihon kodai, chusei-ni okeru kazoku chitsujo – kon'in keisei-to tsuma-no yakuwari nadokara. In *Gender-no Chugokushi*, ed. Obama Masako. Bensei Publishing Inc. (in Japanese).

Barclay, George W., Ansley J. Coale, Michael A. Stoto, and T. James Trussell. 1976. A reassessment of the demography of traditional rural China. *Population Index* 42 (4): 606–635.

Bird, Isabella L. 1898. *Korea and Her Neighbours*. John Murray.

Boserup, E. 1965. *The Conditions of Agricultural Growth: The Economic of Agrarian Change under Population Pressure*. Aldine Publishing Co.

Caldwell, John C. 2001. What do we know about Asian population history? Comparisons of Asian and European research. In *Asian Population History*, ed. Liu Ts'ui-jung, James Lee, David Sven Reher, Saito Osamu, and Wang Feng, 3–23. Oxford University Press.

Campbell, Cameron, and Kurosu Satomi. 2018. Asian historical demography. In *Routledge Handbook of Asian Demography*, ed. Zhao Zhongwei, and Adrian C. Hayes. Routledge.

Cao, Shuji. 1997. Shuyi liuxing-yu huabei shehui-de bianqian: 1580–1644-nian. *Lishi Yanjiu* 1997 (1): 17–32 (in Chinese).

Cao, Shuji. 2001. *Zhongguo Renkoushi 5 Qing Shiqi*. Fudan University Press (in Chinese).

Cha, Myung-Soo. 2009. Josen hugi-eui chulsanryeok, samangryeok mich ingu jeungga: Ne jokpo-e natanan 1700–1899 nyeongan saengmok girok-eur iyonghan yeongu. *Korea Journal of Population Studies* 32 (1): 113–137 in Korean.

Cha, Myung-Soo. 2014. *Gia-wa Gijeok-eui Giweon – Hanguk Gyeongjesa, 1700–2010*. Haeam (in Korean).

Chen, Shao-Sheng. 1979. *Taiwan-de Renkou Bianqian yu Shehui Bianqian*. Lianjing (in Chinese).

Cumings, Bruce. 2005. *Korea's Place in the Sun: A Modern History*, updated ed. W. W. Norton & Co.

Davenport, Romola J. 2020. Urbanization and mortality in Britain, c. 1800–50. *The Economic History Review* 73 (2): 455–485.

Diamond, Jared. 1997. *Guns, Germs, and Steel: The Fates of Human Societies*. W. W. Norton.

Dong, Ke. 2014. Kodai nihon-ni okeru Ijo kisho, toshi seikatsu kankyo-to ekibyo ryuko-ni tsuite. *Journal of East Asian Cultural Interaction Studies* 7: 423–434 (in Japanese).

Eisenstadt, S.N. 1996. *Japanese Civilization: A Comparative View*. University of Chicago Press.

Fagan, Brian. 2004. *The Long Summer: How Climate Changed Civilization*. Basic Books.

Farris, William Wayne. 2006. *Japan's Medieval Population—Famine, Fertility, and Warfare in a Transformative Age*. University of Hawai'i Press.

Fauve-Chamoux, Antionette, and Emiko Ochiai. 2009. Introduction. In *The Stem Family in Eurasian Perspective—Revisiting House Societies, 17th–20th Centuries*, ed. Antoinette Fauve-Chamoux, and Ochiai Emiko, 1–50. Peter Lang.

Fujikawa, Yu. 1912. *Nihon Shippei-shi*. Tofudo (in Japanese).

Fujio, Shin'ichiro. 2015. *Yayoi Jidai-no Rekishi*. Kodansha (in Japanese).

Fukuyama, Francis. 1995. *Trust: The Social Virtues and the Creation of Prosperity*. Free Press.

Glass, D.V. 1965. Population and population movements in England and Wales, 1700 to 1850. In *Population in History*, ed. D.V. Glass, and D.E.C. Eversley, 221–226. Transaction Publishes.

Goode, William J. 1963. *World Revolution and Family Patterns*. The Free Press of Glencoe.

Hajnal, John. 1965. European marriage patterns in perspective. In *Population in History*, ed. D.V. Glass, and D.E.C. Eversley, 101–143. Transaction Publishes.

Hamano, Kiyoshi. 2007. *Kinsei Kyoto-no Rekishi Jinkogaku – Toshi Chonin-no Shakai Kozo-wo Yomu*. Keio University Press (in Japanese).

Harari, Yuval Noah. 2011. *Sapiens: A Brief History of Humankind*. Harvill Secker.

Hayami, Akira. 2015. *Japan's Industrious Revolution: Economic and Social Transformation in the Early Modern Period*. Springer.

Hayami, Akira, and Nobuko Uchida. 1972. Size of household in a Japanese county throughout the Tokugawa era. In *Household and Family in Past Time*, ed. Peter Laslett, 473–515. Cambridge University Press.

Hirai, Shoko. 2008. *Nihon-no Kazoku-to Life Course*. Minerva Shobo (in Japanese).

Ho, Ping-Ti. 1959. *Studies on the Population of China, 1368–1953*. Harvard University Press.

Horiguchi, Tomoichi. 1972. Nihon-no kinsei-ni okeru shippei-no rekishi chirigakuteki kenkyu. *Bulletin of the Faculty of Education, Ibaraki University* 21: 69–79 in Japanese.

Iijima, Wataru. 2009. *Kansensho-no Chugoku-shi – Koshu Eisei-to Asia*. Chuokoronsha (in Japanese).

Imperial Japanese Army General Staff Office. 1924. *Nihon Senshi Chosen-eki*. Murayama Shoten (in Japanese).

Inoue, Kazue. 2005. Chosen koki-ni okeru jinko-to Kazoku-no hen'yo. In *Higashi Asia Kindai Keizai-no Keisei-to Hatten – Higashi Asia Shihonshugi Keiseishi* I, ed. Nakamura Satoru, 89–114. Nihon Hyoronsha (in Japanese).

Institute of Social Engineering. 1974. *Nihon Retto-ni okeru Jinko Bunpo-no Chokiteki Jireiretsuteki Bunseki: Jikeiretsu Suikei-to Yoin Bunseki* (in Japanese).

Ishi, Hiroyuki. 1995. Byoki-no tozai koryux. In *Koza Bunmei-to Kankyo 7. Jinko, Sippei, Saigai*, ed. Hayami Akira, and Machida Hiroshi, 154–165. Asakura Shoten (in Japanese).

Ishi, Hiroyuki. 2018. *Kansensho-no Sekaishi*. Kadokawa (in Japanese).

Ishi, Yoshikuni. 1972. *Kankoku-no Jinko Zoka-no Bunseki*. Keiso Shobo (in Japanese).

Ito, Shigeru. 1995. Kindai nihon-no toshi-ni okeru sippei-to jinko. In *Koza Bunmei-to Kankyo 7. Jinko, Sippei, Saigai*, ed. Hayami Akira, and Machida Hiroshi, 183–196. Asakura Shoten (in Japanese).

Kamada, Motokazu. 1984. Nihon kodai-no jinko-ni tsuite. *Mokkan Kenkyu* 6: 131–154 in Japanese.

Kazama, Kiyozo. 1993. *In'ougo-no Kokyo-wo Saguru*. Iwanami Shoten (in Japanese).

Kikuchi, Isao. 2000. *Kikin – Ue-to Shoku-no Nihonshi*. Shueisha (in Japanese).

Kim, Doo-Sub. 1992. Joseonhugi doshi-e daehan inguhakjeok jeobgeun. *Korean Journal of Sociology* 45: 7–23 in Korean.

Kim, Jae-Ho. 2010a. Joseon hugi hanguk nongeob-eui deukjing-gwa gihusaengtaehakjeok baegyeong. *Asian Comparative Folklore* 41: 97–127 (in Korean).

Kim, Soon-Ja. 2010b. Goryeo shidae-eui jeonjaeng, jeonyeombyeong-gwa ingu. In *Jeonyeombyeong-eui Munhwasa – Goryeo Shidae-reul Boneun Tto Hana-eui Shiseon*, ed. Kim Young-Mi, Lee Hyun Sook, Kim Soon-Ja, Lee Jun-Sook, and Kwon Bok Kyu, 77–114. Hyean (in Korean).

Kim, Keun-Tae. 2016a. 18segi chohon-gwa jaehon-eui sahoesa: Danseong hojeog-eul jungshim-euro. In *Hanguk Yeoksa Inguhak-eui Ganeungseong*, ed. Son Byeong-gyu, 171–199. Sungkyunkwan University Press (in Korean).

Kim, Keun-Tae. 2016b. 18segi doshi-wa nongchon-eui gyeorhon yangsang. In *Hanguk Yeoksa Inguhak-eui Ganeungseong*, ed. Son Byeong-gyu, 200–248. Sungkyunkwan University Press (in Korean).

Kimura, Shigemitsu. 2010. Genshi. In *Nihon Nogyoshi*, ed. Kimura Shigemitsu, 11–38. Yoshikawa Kobunkan (in Japanese).

Kinoshita, Futoshi. 2002. Tokugawa jidai-ni okeru kuraishisuki-no sibo kozo. In *Kindai Ikoki-no Jinko-to Rekishi*, ed. Hayami Akira, 23–43. Minerva Shobo (in Japanese).

Kito, Hiroshi. 1995. Nihon bunmeishi-ni okeru kankyo-to jinko. In *Koza Bunmei-to Kankyo 7. Jinko, Sippei, Saigai*, ed. Hayami Akira, and Machida Hiroshi, 266–279. Asakura Shoten (in Japanese).

Kito, Hiroshi. 2000. *Jinko-kara Yomu Nihon-no Rekishi*. Kodansha (in Japanese).

Kito, Hiroshi. 2001. Jinkoshi-ni okeru kinsei. In *Rekishi Jinkogaku-no Frontier*, ed. Hayami Akira, Kito Hiroshi, and Tomobe Ken'ichi, 45–65. Toyo Keizai Shimposha (in Japanese).

Kito, Hiroshi. 2007. *Zusetsu: Jinko-de Miru Nihonshi – Jomon Jidai–kara Kinmirai Shakai-made*. PHP Interface (in Japanese).

Kobayashi, Fumio. 1942. *Kinsei Shina Keizaishi Kenkyu*. Kobundo (in Japanese).

Korean Society of Infectious Diseases (KSID). 2009. *Hanguk Gamyeombyeongsa*. Koonja Publishing Inc. (in Korean).

Kurachi, Katsunao. 2016. *Edo-no Saigai-shi – Tokugawa Nihon-no Keiken-ni Manabu*. Chuko Shinsho (in Japanese).

Kurosu, Satomi. 2009. Marriage, divorce and remarriage in a stem-family system: Women in two northeastern Japanese villages, 1716–1870. In *The Stem Family in Eurasian Perspective—Revisiting House Societies, 17th–20th Centuries*, ed. Antoinette Fauve-Chamoux, and Ochiai Emiko, 327–344. Peter Lang.

Kurosu, Satomi, Tsuya Noriko, and Hamano Kiyoshi. 2012. Tokugawa-ki kohan-ni okeru shokon pattern-no chiikisa. In *Rekishi Jinkogaku-kara Mita Kekkon, Rikon, Saikon*, ed. Kurosu Satomi, 24–56. Reitaku University Press (in Japanese).

Kuwano, Eiji. 2002. Chosen shakai-no sangyo kozo. In *Kankokugaku-no Subete*, ed. Furuta Hiroshi, and Ogura Kizo, 104–107. Shinshokan (in Japanese).

Kwon, Tai-Hwan, and Kim Doo-Sub. 2002. *Ingu-eui Ihae*. Seoul National University Press (in Korean).

Kwon, Tai-Hwan., and Sin Yongha. 1977. Josen wangjo shidae ingu chugye-eui gwanhan ilshiron. *Donga Munhwa* 14: 289–330 in Korean.

Kwon, Tai-Hwan, Lee Hae-Young, Chang Yunshik, and Yu Eui-Young. 1975. *The Population of Korea*. Population and Development Studies Center, Seoul National University.

Kye, Bong-Oh, and Park Hee-Jin. 2016. 20-segi choban Seoul-eui yeonryeong-byeol ingu idong pattern yeongu. In *Hanguk Yeoksa Inguhak-eui Ganeungseong*, ed. Son Byeong-gyu, 359–402. Sungkyunkwan University Press (in Korean).

Laslett, Peter. 1972. Introduction: The history of the family. In *Household and Family in Past Time*, ed. Peter Laslett, 1–89. Cambridge University Press.

Lee, Hunchang. 2004. *Kankoku Keizai Tsuushi*. Translated by Sugawa Hidenori, and Rokutanda Yutaka. Hosei University Press (in Japanese).

Lee, Hyun Sook. 2010. Jeonyeombyeong, chiryo, gweongryeog – goryeo shidae jenyeombyeong-eui yuhaeng-gwa chiryo. In *Jeonyeombyeong-eui Munhwasa – Goryeo Shidae-reulFe Boneun Tto Hana-eui Shiseon*, ed. Kim Young-Mi, Lee Hyun Sook, Kim Soon-Ja, Lee Jun-Sook, and Kwon Bok Kyu, 19–76. Hyean (in Korean).

Lee, James, and Cameron Campbell. 1997. *Fate and Fortune in Rural China: Social Organization and Population Behavior in Liaoning 1774–1873*. Cambridge University Press.

Lee, James J., and Wang Feng. 1999. *One Quarter ofHumanity: Malthusian Mythology and Chinese Realities 1700–2000*. Harvard University Press.

Liu, Ts'ui-Jung, and Shi-Yung Liu. 2001. Disease and mortality in the history of Taiwan. In *Asian Population History*, ed. Liu Ts'ui-jung, James Lee, David Sven Reher, Saito Osamu, and Wang Feng, 248–269. Oxford University Press.

Livi-Bacci, Massimo. 2012. *A Concise History of World Population*, 5th ed. Wiley.

Maddison, A. 2001. *The World Economy: A Millennial Perspective*. Development Centre of the OECD.

Malthus, T. R. 1798. *An Essay on the Principle of Population*.

McEvedy, Colin, and Richard Jones. 1978. *Atlas of World Population History*. Penguin Reference Books.

McNeill, William H. 1976. *Plagues and Peoples*. Anchor.

Michell, Tony. 1979/80. Facts and hypothesis in Yi Dynasty economic history: The demographic dimension. *Korea Studies Forum* 6: 65–93.

Minegishi, Sumio. 2011. *Chusei Saigai, Senran-no Shakaishi*. Yoshikawa Kobunkan (in Japanese).

Miyajima, Hiroshi. 1994. Higashi Asia shono shakai-no keisei. In *Choki Shakai Hendo: Asia-kara Kangaeru*, ed. Mizoguchi Yuzo, Hamashita Takeshi, Hiraishi Naoaki, and Miyajima Hiroshi, 67–96. Tokyo University Press (in Japanese).

Mizuno, Shoji. 2006. Saigai-to kankyo-heno shiten. In *Nihon Saigaishi*, ed. Kitahara Itoko, 86–96. Yoshikawa Kobunkan (in Japanese).

Nakamura, Satoru. 2005. Higashi Asia sihonshugi keiseishi josetsu. In *Higashi Asia Kindai Keizai-no Keisei-to Hatten – Higashi Asia Shihonshugi Keiseishi* I, ed. Nakamura Satoru, 1–20. Nihon Hyoronsha (in Japanese).

Nakane, Chie. 1970. *Japanese Society*. University of California Press.

Nakane, Chie. 1972. An interpretation of the size and structure of the household in Japan over three centuries. In *Household and Family in Past Time*, ed. Peter Laslett, 517–543. Cambridge University Press.

National Population and Family Planning Commission. 2011. *Renkou Yu Jihua Shengyu Changyong shuju shouce 2011* (in Chinese).

Ochiai, Emiko. 2009. Two types of stem household system in Japan: the Ie in global perspective. In *The Stem Family in Eurasian Perspective—Revisiting House Societies, 17th–20th Centuries*, ed. Antoinette Fauve-Chamoux, and Ochiai Emiko, 287–326. Peter Lang.

Okamoto, Takashi. 2013a. Chugoku-no keizai-to rekishi. In *Chugoku Keizaishi*, ed. Okamoto Takashi, 177–236. Nagoya University Press (in Japanese).
Okamoto, Takashi. 2013b. Minshin – dento keizai-no keisei-to hensen (15–19 Seiki). In *Chugoku Keizaishi*, ed. Okamoto Takashi, 177–236. Nagoya University Press (in Japanese).
Okamoto, Takashi. 2015. *Kindai Chugoku-shi*. Chikuma Shinsho (in Japanese).
Okuno, Masao. 2000. *Tetsu-no Kodaishi 3 – Kibabunka*. Hakusuisha (in Japanese).
Park, Hee-Jin. 2016. Joseon shide yanbang-eui honin yeonryeong-gwa jaehon geumji-ga chulsan-e michineun yeonhyang – munjib, jokbo-reul jungshim-euro. In *Hanguk Yeoksa Inguhak-eui Ganeungseong*, ed. Son Byeong-gyu, 249–267. Sungkyunkwan University Press (in Korean).
Park, Hee-Jin, and Myung-Soo Cha. 2004. Joseonhugi-wa iljeshide-eui ingubyeondong, Jeonjuyissi jangcheongunpa-wa hamyangbakssi jeongranggongpa jokbo-eui bunseok. In *Suryanggyeongjesa-ro Dashi Bon Joseonhugi*, ed. Lee Young-Hoon, 1–40. Seoul National University Press (in Korean).
Saito, Osamu. 2002. Dento chugoku-no rekishi jingogaku. *Shakai Keizai Shigaku* 68 (2): 87–99 in Japanese.
Sakane, Yoshihiro. 2010. Kindai. In *Nihon Nogyoshi*, ed. Kimura Shigemitsu, 255–336. Kikkawa Kobunkan (in Japanese).
Sato, Yoichiro. 2002. *Ine-no Nihonshi* (in Japanese). Kadokawa Sophia Bunko.
Savage-Landor, A.H. 1895. *Corea or Cho-sen: The Land of the Morning Calm*. William Heineman.
Sawada, Goichi. 1927. *Naracho Jidai Minsei Keizai-no Suteki Kenkyu*. Fuzanbo (in Japanese).
Shikata, Hiroshi. 1976. *Chosen Shakai Keizaishi Kenkyu Chu*. Kokusho Kankokai (in Japanese).
Shima, Mutsuhiko. 2004. Chokiteki shiya-ni okeru kankoku-no kazoku (in Japanese). In *Henbo-suru Higashi Asia-no Kazoku*, eds. Sato Yasuyuki, Shimizu Hiroaki, and Kisaki Tetsuro, 81–109. Waseda Universtiy Press.
Suzuki, Toru. 2013. *Low Fertility and Population Aging in Japan and Eastern Asia*. Springer.
Suzuki, Toru. 2019. *Eastern Asian Population History and Contemporary Population Issues*. Springer.
Tange, Yasushi. 2014. *Ijo Kisho-ga Kaeta Jinrui-no Rekishi*. Nihon Keizai Shinbunsha (in Japanese).
Tetsuyama, Hiroshi. 2000. Chugoku dento nogyo keizai-to sono kindaiteki henyo. In *Higashi Asia Keizaishi-no Shomondai*, ed. Matsuda Koichi, 17–37. Aunsha (in Japanese).
Tomobe, Ken'ichi. 2001. The level of fertility in Tokugawa and Meiji Japan, C. 1800s–1930s: A preliminary analysis of the Hutterite indices. In *Asian Population History*, ed. Liu Ts'ui-jung, James Lee, David Sven Reher, Osamu Saito, and Wang Feng, 138–151. Oxford University Press.
Tsuji, Yoshinobu. 2006. Nihon-ni okeru rekisi tsunami. In *Nihon Saigai-shi*, ed. Kitahara Itoko, 261–264. Yoshikawa Kobunkan (in Japanese).
Tsuya, Noriko O. 2001a. Patterns of fertility and nuptiality in a fishing village in southwestern Tokugawa Japan. In *Asian Population History*, ed. Liu Ts'ui-jung, James Lee, David Sven Reher, Saito Osamu, and Wang Feng, 107–137. Oxford University Press.
Tsuya, Noriko. 2001b. Kinsei nihon-no shussei regime – Oshu Nihonmatsuhan-no ninbetsu aratamecho data-no event history bunseki. In *Rekishi Jinkogaku-no Frontier*, ed. Hayami Akira, Kito Hiroshi, and Tomobe Ken'ichi, 219–244. Toyo Keizai Shimposha (in Japanese).
Tsuya, Noriko. 2002. Kinsei koki noson-ni okeru jinko zoka-to shusseiryoku-no bunseki – hizen-no-kuni sonogi-gun nomo-mura-no jirei. In *Koza Bunmei-to Kankyo 7. Kindai Ikoki-no Jinko-to Rekishi*, ed. Hayami Akira, 175–198. Minerva Shobo (in Japanese).
Umezao, Tadao. 1967. *Bunmei-no Seitai Shikan*. Chuo Koronsha (in Japanese).
Wakabayashi, Masahiro. 2001. *Taiwan – Hen'yo-shi Chucho-suru Identity*. Chikuma Shobo (in Japanese).
Wakabayashi, Keiko, and Haisong Nie. 2012. *Chugoku Jinkou Mondai-no Nenpu-to Tokei: 1949–2012-nen*. Ochanomizu Shobo (in Japanese).
Wall, Richard. 1995. Historical development of the household in Europe. In *Household Demography and Household Modeling*, ed. E. van Imhoff, A. Kuijsten, P. Hooimeijer, and L. van Wissen, 19–52. Plenum Press.

Wilson, Chris. 2001. Understanding the nature and importance of low-growth demographic regimes. In *Asian Population History*, ed. Liu Ts'ui-jung, James Lee, David Sven Reher, Osamu Saito, and Wang Feng, 24–44. Oxford University Press.

Wittfogel, K.A. 1959. *Oriental Despotism.* Yale University Press.

Wolf, Arthur P., and Chieh-shan Huang. 1980. *Marriage and Adoption in China, 1845–1945.* Stanford University Press.

Woo, Seungyeop. 2016. *Daegigeun-i Onda – Sesang-eul Bakkun Gamum-gwa Gigeun-eui Yeoksa.* Cheom Books (in Korean).

Yasuda, Yoshinori. 1995. Pesuto dairyuko (in Japanese). In *Koza Bunmei-to Kankyo 7. Jinko, Sippei, Saigai*, ed. Hayami Akira, and Machida Hiroshi, 120–132. Asakura Shoten.

Yi, Hyong-nang. 2015. *Shokuminchi Chosen-no Kome-to Nihon.* Chuo University Press (in Japanese).

Zensho, Eisuke. 1927. *Chosen-no Jinko Gensho.* Governor-General of Korea (in Japanese).

Zhao, Gang, and Zhong-Yi Chen. 1989. *Zhongguo Nongye Jingjihi.* Youshi Wenhua Shiye Gongsi (in Chinese).

Chapter 5
Population in Modern Eastern Asia

5.1 Western Impacts

In Eastern Asia, historical incidents that mark the beginning of the modern period are the First Opium War (1840–42) for China, the Meiji Restoration (1868) for Japan, the Ganghwa Island Incident (1875) for Korea, and the Annexation to Japan (1895) for Taiwan. All these incidents were affected by Western impacts, although the latter two were indirectly transmitted by Japan. The transition from the modern to the contemporary periods are marked by Japan's defeat in the Pacific War in 1945 for Japan, Korea, and Taiwan, and by the establishment of the People's Republic of China in 1949 for China.

In the 1840s, the military power of European nations, especially the potency of British navy, was already strong enough to enforce their economic concerns in China. The British had practically colonized India by the early nineteenth century. The First Opium War broke out to secure the British interest in China. After a series of defeats on the Changjiang River, the Qing dynasty had to conclude the Treaty of Nanking in 1842, opening five ports and abandoning Hong Kong. Although the Qing dynasty survived until 1911, the political and economic sovereign of China was repeatedly challenged by Western powers and Japan.

The Japanese in the early nineteenth century were more aware of the Western impact than the Chinese. The Japanese intellectuals learned the European medicine and other sciences, and the shogunate officials were informed the world affairs from the Dutch stayed in Nagasaki. When the US commodore M. C. Perry came to Japan to open the country, the shogunate knew that Japan could not resist long. When Japan concluded the Convention of Kanagawa in 1854, anti-shogun feudal domains such as Satsuma and Choshu devoted themselves to the "sonno joi" movement, revering the emperor and expelling foreigners. However, they quickly learned the necessity of accepting Western institutions and technologies shortly after the battles against the Western powers. After a short civil war in 1868, the Meiji government was established by the leaders of Satsuma and Choshu. Under the slogan of "rich country

T. Suzuki, *Comparative Population History of Eastern Asia*,
https://doi.org/10.1007/978-981-99-9367-3_5

and strong army," Japan quickly introduced the Western institutions and cultures. As discussed above, Japan successfully achieved this task because the Japanese society was already similar to Europe.

While China and Japan opened to the West, Korea remained to be a "hermit nation," refusing the trade with the Western countries. Although France and the US attacked Korea in 1866 and 1871, respectively, these expeditions were not successful. It was Japan that forced the Yi Joseon dynasty to open the country in 1875. After a battle between the Ganghwa Island fort and the Japanese gunboat Un'yo, the Yi dynasty agreed with the Treaty of Ganghwa, opening three ports to Japan. Korea finally gave up its position of hermit nation in 1882, concluding treaties with Western countries including the US, the UK and Germany.

Japan attempted to separate Korea from Chinese influence to seek its own interests on the peninsula. This caused a serious conflict and resulted in the First Sino-Japanese War in 1894–95. Japan pressured China both on land and sea. The Shimonoseki Treaty was a decision to cede Taiwan and the Liaodong Peninsula to Japan. However, because of the Triple Intervention of Russia, France and Germany, Japan was forced to give up Liaodong. The Japanese rule of Taiwan lasted for 50 years until Japan surrendered to the Allies in 1945.

After China was pushed out of the Korean peninsula, Russia increased its influence instead. When Queen Min was assassinated by a group of Japanese, King Gojong (reign 1863–1907) took refuge at the Russian legation in 1896. Although the king returned to the palace in 1897 and proclaimed himself to be the first emperor of the Great Korean Empire, Japan and Russia continued the contest to be the actual owner of the peninsula. When all diplomatic negotiations failed, the Russo-Japanese War broke out in 1904. Japan managed to occupy Port Arthur and defeated the Baltic Fleet on the Sea of Japan. The Treaty of Portsmouth concluded in 1905 authorized Japan's paramount interest in the Korean peninsula. Japan finally annexed Korea in 1910 and its colonial rule lasted for 35 years.

5.2 Modern Population Growth in Japan

Protoindustrialization in Japan began in the early nineteenth century at the end of the premodern period. It is assumed that the feudal lords' efforts to stimulate the local economy, the inflation caused by coin-minting by the Tokugawa shogunate, increased agricultural productivity induced the growth of manufactural industry in rural areas (Ishii 1991; Kito 2007). This progress was directly linked to the industrialization and modern population growth that occurred after Japan opened to the West and was built into the global economy. Japan had to achieve the rapid industrialization to catch up with the West. While the production of goods that could make immediate profits grew rapidly, the infrastructural developments were left behind. Because Japan embarked on modernization in the age of the imperialist race for colonial expansion, an oversized military power was required and sacrificed the living standard of people (Hayami 2015).

Table 5.1 Population of modern Japan

Year	Population (thousand)	Year	Annual growth (%)
1872	34,806	1872–1880	0.65
1880	36,649	1880–1885	0.89
1885	38,313	1885–1890	0.82
1890	39,902	1890–1895	0.82
1895	41,557	1895–1900	1.08
1900	43,847	1900–1905	1.23
1905	46,620	1905–1910	1.08
1910	49,184	1910–1915	1.41
1915	52,752	1915–1920	1.01
1920	55,473	1920–1925	1.49
1925	59,737	1925–1930	1.53
1930	64,450	1930–1935	1.45
1935	69,254	1935–1940	0.76
1940	71,933	1940–1945	0.06
1945	72,147		

Based on IPSS (2023)

In 1872, the Meiji government conducted a comprehensive survey to establish the population register. Populations in Table 5.1 before the first modern census in 1920 are estimates by the Statistics Office based on the population register. While the annual growth rate was less than 1% in the nineteenth century, the population growth accelerated in the twentieth century. There were decelerations due to the depression in 1907 and the influenza pandemic in 1918–20. The growth rate peaked in the late 1920s. Although fertility declined to some extent since the 1920s, the main response to population pressure was emigration. When the US, Canada, and Australia closed the gate to the Chinese in the late 1880s and 1890s, Japanese migrants became the mainstream from Asia. After the turn of the century, however, the Japanese migrants were also restricted in the US and finally banned in 1924. While Brazil also constrained Japanese migrants in 1934, the number of overseas Japanese in Korea, Manchuria, mainland China, and Southeastern Asia kept growing (Shiode 2015).

5.3 Modern Population Growth in Korea

It is apparent that the population of Korea increased during the Japanese era. However, the population size before the first modern census in 1925 is not clear. Because the coverage of the population register was improved markedly in the early 1910s, the growth rate based on these official populations would be exaggerated. As shown in Table 5.2, the estimated population of 1910 varies from 15.5 million (Ishi 1972) to

17.4 million (Kwon et al. 1975). Ishi Yoshikuni (aka Seok Nam-Guk) estimated the population at the beginning of the year, while others accepted census populations on October 1 of 1925, 1930, 1935, and 1944 without adjustment. Thus, Ishi's estimates are slightly smaller than census results.

The wide range of estimated Korean populations derived from the fact that the population register before the census was so incomplete that demographers were forced assumptions without empirical basis. Kwon Tai-Hwan and his colleagues assumed that the natural growth rate was 0.2% in the 1900s and that it accelerated to 1.87% in the 1920s without any empirical support. Kim Cheol conducted a backward projection from 1925 by applying assumed natural growth rates and the number of emigrants. Park (2009) used basically the same method but applied slightly lower natural growth rates than Kim. Cha (2014) estimated crude birth and death rates based on his analysis of genealogical records and applied those rates to the backward projection. However, it is difficult to apply the finding of male yangbans to the total population. He estimated adult mortality assuming that the differential mortality by class before 1925 was same as what observed for 1925–40. Even if this assumption were correct, the anxiety remains that the genealogical records that he analyzed had serious sampling bias. Thus, it is difficult to say that the introduction of genealogical analysis significantly improved the backward population projection in the Japanese era.

Table 5.2 Estimations for Korean population in Japanese era

Year	Official	Ishi (1972)	Kim (1965)	Cha (2014)	Park (2009)	Park (2008)	Kwon et al. (1975)
Population (1000)							
1910	13,129	15,474	16,310	16,272	16,541	17,001	17,427
1915	15,958	16,485	17,027	17,566	17,327	17,823	17,656
1920	16,916	17,533	17,629	18,296	17,854	18,073	18,072
1925	19,020	18,797	19,020	19,020	19,020	19,020	19,020
1930	20,438	20,219	20,438	20,438	20,438	20,438	20,438
1935	22,208	21,890	22,208	22,208	22,208	22,208	22,208
1940	23,547	23,342	23,547	23,547	23,547	23,547	23,547
1944	25,120	24,917	25,120		25,120		
Average annual growth rate (%)							
1910–15	3.98	1.27	0.86	1.54	0.93	0.95	0.26
1915–20	1.17	1.24	0.70	0.82	0.60	0.28	0.47
1920–25	2.37	1.40	1.53	0.78	1.27	1.03	1.03
1925–30	1.45	1.47	1.45	1.45	1.45	1.45	1.45
1930–35	1.67	1.60	1.67	1.67	1.67	1.67	1.67
1935–40	1.18	1.29	1.18	1.18	1.18	1.18	1.18
1940–44	1.63	1.65	1.63		1.63		

Ishi Yoshikuni and Park Yi-Taek projected not only the total population but also the population by sex and five-year age group. Ishi's backward population projection relied on his own estimation of life tables for 1905–10, 1910–15, 1915–20, and 1920–25. According to his life table for 1905–10, mortality was very high and the life expectancy at birth was 27.7 and 29.0 for male and female, respectively. These values are not necessarily unrealistic considering that the life expectancy at birth in rural China around 1930 was 24.6 and 23.7 for male and female, respectively (Barclay et al. 1976). On the other hand, Park Yi-Taek criticized that the number of deaths implied by Ishi exceeded the number of registered death certificates. Although the annual number of registered deaths could temporarily exceed the actual number if the registration concentrated in a specific year, it is unnatural that such over-registration continued for years. Thus, morality in Ishi's life table should be considered as an overestimation, which implies that the population growth until 1925 was also an exaggeration.

While Park Yi-Taek's criticism was convincing, his own projection was not necessarily more trustworthy than Ishi. For example, Park estimated the coverage of death register in 1926–40 based on the number of emigrants estimated by Kim (1965). However, Kim's assumption that mortality of Koreans in Manchuria was same as those remained in Korea is questionable if migrants were younger than those in homeland. Furthermore, Park estimated coverage of death registration in 1910–24 assuming that the population register in 1914 was correct. Because the registered population was considerably smaller than the actual population, there should have been more deaths than Park's calculation. Then, it is deduced that the actual population in 1910 was somewhere between estimations by Ishi and Park. This is why the total population of 16–17 million was assumed in Table 4.11. This implies that the average annual growth rate in 1910–25 was between 0.75 and 1.16%, which was lower than 1.30% of Japan in the same period. The impact of influenza pandemic was small in results of Ishi and Kim Cheol, and the population growth even accelerated according to Kwon. Considering the case of Japan, results of Cha Myung-Soo, Park Keong-Suk, and Park Yi-Taek are more realistic. The Korean population began to grow faster than the Japanese after 1930, which accelerated emigration to Japan, Manchuria, China, Sakhalin, and other regions.

5.4 Modern Population Growth in Taiwan

After the first census to establish the Taiwanese population register in 1905, Japan conducted more modernized censuses in every fifth year between 1915 and 1940. The completeness and accuracy of these censuses were well above those in the Western world (Barclay 1954). Mizoguchi (2008) interpolated populations between censuses and extrapolated those in 1898–1904 and 1941–43 by applying growth rates of the population registry. He added the result of the population survey on February 22, 1944, to his table, while other populations are on October 1st of each year.

All the population figures for Taiwan in Table 5.3 were taken from Table 2.1 of Migoguchi (2008). According to the table, the population of Taiwan increased more rapidly than Japan and Korea. The average annual growth rate in 1910–44 was 1.95%, which was higher than those of Japan and Korea. In fact, Taiwan's population growth rate was constantly over 2% in 1924–42. The rapid population growth in Taiwan implies a low developmental level before the annexation and Japan's successful policy intervention to reduce mortality. Taiwan was notorious until the nineteenth century as a very unhealthy island. Troops of Koxinga, the Qing dynasty, France, and Japan that landed in Taiwan suffered serious damages caused by malaria and other infectious diseases (Liu and Liu 2001). When the Japanese army occupied Taiwan in 1895, only 154 soldiers died in combat but more than 4000 died of disease within several months. The Governor-General's office eradicated plague and cholera through strict vigilance, improved the water supply system, and controlled malaria through blood tests and treatment with quinine (Barclay 1954). Unlike in UK-ruled India where mortality of Indians did not decline, Japanese officials enforced the same public health measures in Japanese and Taiwanese communities in Taiwan (Wakimura 2002).

This successful intervention by Japan was supported by the "paojia" system; an organization of households that already existed under the rule of the Qin dynasty. Paojias were built into the police system and helped police officers in reporting and segregating diseases, avoiding contamination of the water supply, monitoring food handling and butchery, and managing burials and cremations. The elaborately organized police and paojia systems contributed to the great accuracy of population registry and vital statistics. Barclay (1954) found that the consistency between the number of survivors in censuses and the counts of births and deaths in vital statistics were almost perfect for the period between 1905 and 1920. Thus, Japan's management in Taiwan was far more successful not only than Korea but also than most areas colonized by Western powers (Cumings 1997).

Table 5.3 Population of modern Taiwan

Year	Population (thousand)	Year	Annual growth (%)
1900	2,807	1900–1905	1.87
1905	3,080	1905–1910	1.11
1910	3,254	1910–1915	1.59
1915	3,520	1915–1920	0.97
1920	3,694	1920–1925	1.95
1925	4,067	1925–1930	2.46
1930	4,593	1930–1935	2.56
1935	5,212	1935–1940	2.41
1940	5,872	1940–1945	1.32
1945	6,270		

Mizoguchi (2008)

5.5 Population of Modern China

Table 5.4 summarizes population estimates in modern China including the late Qing and the Republic of China periods. Kobayashi (1942) accepted existing estimates for 1885, 1894, and 1901 as valid figures. According to Zhao and Xie (1988), as cited by Ueda (1995), the population decreased by 71.7 million (16.7%) in 1855–75. Cao (2001), as cited in Mizushima et al. (2015), also estimated that the population decreased by 72 million (16.5%) in 1850–80. While Maddison's estimate for 1870 is consistent with others, his result for 1885 seems to be unnaturally high. The population during the Republic of China period was taken from Minami and Makino (2014).

Figure 5.1 compares average annual growth rates in Eastern Asia in 1700–1945 with England and Wales. The average of six different growth rates shown in Obuchi and Morioka (1981) were used for England and Wales before 1801. Growth rates in the nineteenth and twentieth centuries were calculated from census results. For China, estimates by Kobayashi (1942), Zhao and Xie (1988), Minami and Makino (2014) were chosen to calculate annual growth rates. Rates for Japan were obtained from Tables 4.4 and 5.1. For Korea, while the growth rates in the nineteenth century

Table 5.4 Population estimates of modern China (in million)

Year	Kobayashi (1942)	Zhao and Xie (1988)	Cao (2001)	Maddison (2001)	Minami and Makino (2014)
1845		428			
1850	397		436		
1855		430			
1865		370			
1870				358	
1875		359			
1880			364		
1885	382	373		426	
1894	421				
1901	426				
1910			436		
1913				437	442
1915					448
1920					462
1925					477
1930					498
1935					532
1940					540
1944					547

were based on the average of low and high variants in Table 4.11, the average of five studies other than Kwon et al. (1975) in Table 5.2 was used to calculate the rates during the Japanese era. For Taiwan, Chen's estimates in Table 4.13 were used to obtain the growth rates in the nineteenth century.

The population growth of China apparently preceded England and Wales and the growth rate was close to 1% as early as the 1730s. As discussed above, China was the first region to take the opportunity of new epidemiological balance achieved by the end of the seventeenth century. Most infectious diseases became childhood diseases that everyone contracted, leading people acquire immunity to them (McNeill 1976).

However, unlike Western countries that experienced the modern population growth in the nineteenth century, China could not sustain simultaneous growth of the population and the economy. Population growth slowed down in the early nineteenth century and finally turned negative in the 1850s. The magnificent population growth of the eighteenth century, with the background of new epidemiological balance and peaceful political climate, was achieved through the agricultural growth planting of maize, potatoes, sweet potatoes and peanuts (Ho 1959; Chen 1979). The extension of farmlands, however, was approaching the limit in the nineteenth century. As population pressure intensified, per capita food consumption decreased, and even a small decline in the harvest resulted in famine. Increasing poverty and social disorder added to the spread of opium addiction reduced labor productivity and accelerated economic decline. Finally, there was a critical population decrease during subsequent

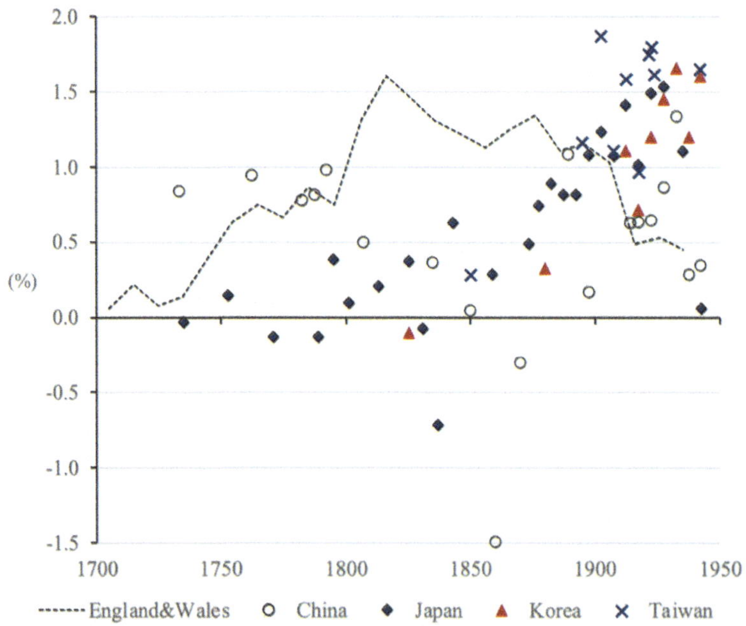

Fig. 5.1 Population growth rate in 1700–1945

riots including the Taiping Rebellion (1850–64), the Nian Rebellion (1851–68), and the Dungan Revolt (1862–77).

Although China took the opportunity afforded by the new epidemiological balance to start the population increase, the fundamental transformation from an agrarian to an industrial society did not happen in the nineteenth century. According to Diamond (1997), misguided policies in the Ming Dynasty shut down the Age of Exploration and discouraged the technological innovation. On the other hand, the diversity in Europe induced technological development through conflict and competition among nations. By the eighteenth century, European countries including England had accumulated enough technological assets to enable the fundamental transition to an industrial society.

In contrast to China, Japan's protoindustrialization began a few decades before the Meiji Restoration. When Japan opened the country and was integrated into the global economy, many Western countries had already exemplified the English model. Japan succeeded in connecting protoindustrialization to genuine industrialization in the late nineteenth century. The population started growing at a rate that could not be sustained in an agrarian society. By the early twentieth century, modern population growth also took place in Korea and Taiwan. China was left behind since its protoindustrialization began too early to learn from England, and its technological accumulation was insufficient to set off industrialization spontaneously. China's failure to synchronize economic and demographic developments appeared again in the late twentieth century.

5.6 Internal Migration in Modern Period

Industrialization is the shift of the labor force from the primary to the secondary and tertiary sectors and always accompanies the urbanization. Cities also attracted young people with educational opportunities. However, not all surplus population in rural areas moved to cities within the country; some migrated to remote frontiers or even emigrated abroad. Zelinsky (1971) asserted in his hypothesis of mobility transition that international and frontier migrations peaked earlier than rural–urban migration. While his statement was deduced from the experience in the Western countries, it also matches with the experience of the Japanese Empire. As shown in Fig. 5.2, the Gini coefficient of population distribution in Japan started rising based on urbanization only after 1920. Until then, the population distribution was rather equalized because of the flow into the frontier of Hokkaido. The equalization in Korea and Taiwan suggests that the impact of urbanization was far from dominant during the Japanese rule.

The Gini coefficient based on provinces of China cannot capture the trend of urbanization but is closer to observing the distributional change of the European population. As a fact, China is too large to be integrated with only one center. Because multiple centers such as Beijing, Shanghai, Tianjin, Chongqing, and Shenyang have had outstandingly large population, the rank-size rule have not worked in China. In

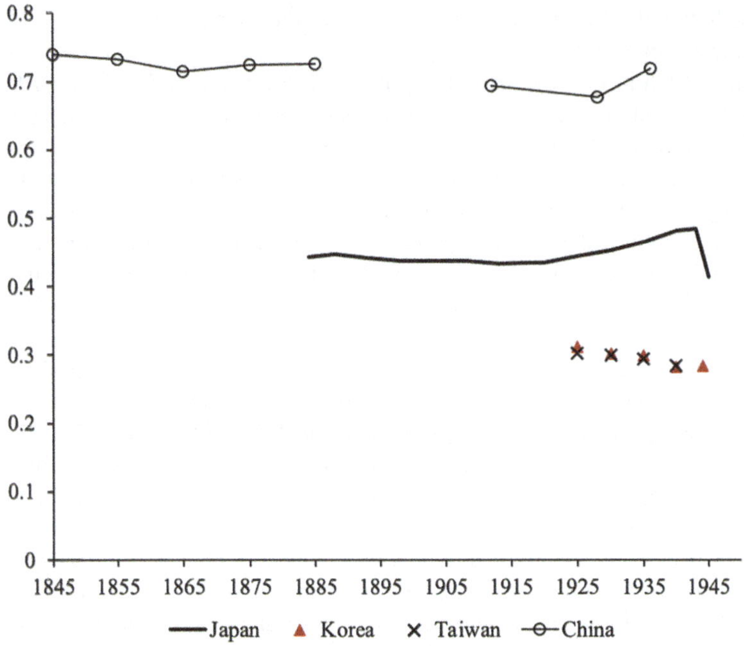

Fig. 5.2 Gini coefficient of population distribution in Eastern Asia: 1845–1945. Based on regional population data by Statistical Office of Japan, Censuses, Zhao and Xie (1988) cited in Ueda (1995), Minami and Makino (2014)

any case, population distributional changes in China until 1949 could be seen as the premodern phase since the modern population increase was not evolving at that time.

Table 5.5 shows population growth rates by prefecture of Japan in the modern era. Until 1920, Hokkaido sustained the highest population growth rate. In the Edo period, there were less than 100,000 Japanese and Ainu population in Hokkaido. The Meiji government sent ex-samurais as "tondenhei," military settler colonists, to develop and defend Hokkaido since 1874. Later, the government encouraged commoners' settlement and many peasants from Tohoku and Hokuriku regions migrated to Hokkaido. By the end of the 1910s, however, the attractiveness of the colonization frontier almost disappeared, and the population growth of Hokkaido slowed down. Instead, urbanized prefectures such as Tokyo, Osaka, and Kanagawa occupied the highest ranking in growth rates. As a result, the Gini coefficient based on prefecture rose from 0.4355 in 1920 to 0.4848 in 1943. In the later stage of the Pacific War, however, there was a huge evacuation flow of urban residents to rural areas as airstrikes by the US military force intensified. Although the process of urbanization was disrupted temporarily by the war, it restarted in the 1950s.

As discussed above, Edo in the eighteenth century had more than one million population and was the most populous city in the world. After the city was renamed Tokyo in 1868, the population increased from 1.39 million in 1889 to 2.17 million

Table 5.5 Population growth rate (%) by prefecture of Japan: 1884–1945

Prefecture	1884–88	1888–93	1893–98	1898–03	1903–08	1908–13	1913–18	1918–20	1920–25	1925–30	1930–35	1935–40	1940–45
Japan total	1.42	0.87	1.11	1.28	1.15	1.42	1.02	0.27	1.31	1.53	1.45	0.76	0.02
Hokkaido	7.87	8.60	10.47	5.33	5.87	4.53	4.42	7.31	1.16	2.39	1.76	1.03	1.73
Aomori	1.83	1.29	1.44	1.44	1.49	1.35	1.03	– 1.92	1.45	1.60	1.91	0.37	1.92
Iwate	1.67	0.98	0.59	0.74	0.68	1.69	0.47	– 0.10	1.28	1.61	1.40	0.61	2.63
Miyagi	3.39	1.44	1.03	1.14	– 0.59	1.34	0.48	1.44	1.66	1.82	1.56	0.19	3.24
Akita	1.71	1.01	1.19	1.40	1.23	1.10	0.69	– 2.79	0.83	1.07	0.99	– 0.05	3.21
Yamagata	1.33	0.92	0.94	1.14	0.69	1.28	0.43	– 0.12	1.18	1.01	0.67	– 0.31	3.82
Fukushima	1.69	1.54	1.46	1.33	0.87	1.27	1.23	0.85	1.08	0.96	0.96	0.18	4.17
Ibaraki	1.76	1.07	1.29	0.83	0.90	1.18	1.07	– 0.37	0.85	1.08	0.82	0.59	4.04
Tochigi	1.98	1.54	1.75	1.87	1.26	1.45	1.16	– 0.40	0.83	0.92	0.92	– 0.14	5.44
Gunma	2.41	1.24	1.26	1.73	1.19	1.66	1.20	0.37	1.23	1.17	0.93	0.59	3.85
Saitama	1.55	0.94	0.71	1.05	0.61	1.19	0.67	– 1.09	1.11	0.91	0.94	0.70	5.28
Chiba	1.14	0.67	0.66	0.66	0.50	0.83	– 0.08	– 0.96	0.93	0.99	1.02	0.19	4.73
Tokyo	4.12	3.50	3.14	3.69	3.56	0.94	3.52	5.24	3.93	3.82	3.33	2.72	– 13.69
Kanagawa	2.79	– 3.74	2.59	2.53	2.23	0.95	1.71	3.03	1.37	2.71	2.58	3.24	– 2.87
Niigata	1.23	0.52	0.01	0.29	0.38	1.24	– 0.11	– 2.47	0.81	0.89	0.64	0.27	3.39
Toyama	1.66	0.68	– 0.49	0.17	– 0.35	0.93	– 0.02	– 3.85	0.68	0.78	0.51	0.28	3.32
Ishikawa	0.35	0.21	– 0.50	0.35	0.18	0.72	– 0.24	– 1.47	0.09	0.16	0.30	– 0.60	3.55
Fukui	0.60	0.39	0.00	0.45	– 0.25	0.80	– 0.45	– 1.92	– 0.04	0.67	0.91	– 0.35	2.67
Yamanashi	1.68	1.07	0.94	1.28	1.05	1.27	0.72	– 1.97	0.58	0.99	0.49	0.13	5.21
Nagano	1.58	1.07	0.85	1.30	0.79	1.39	1.05	1.19	0.84	1.06	– 0.04	– 0.36	4.73
Gifu	1.00	0.63	0.35	0.84	0.10	1.37	0.45	– 0.88	1.14	0.80	0.79	0.28	4.09

(continued)

Table 5.5 (continued)

Prefecture	1884–88	1888–93	1893–98	1898–03	1903–08	1908–13	1913–18	1918–20	1920–25	1925–30	1930–35	1935–40	1940–45
Shizuoka	1.81	1.07	0.95	1.50	1.17	2.01	1.22	– 0.10	1.51	1.47	1.53	0.44	2.29
Aichi	1.29	0.80	0.79	1.33	1.47	1.90	0.95	0.79	2.11	2.05	2.20	1.73	– 1.74
Mie	0.94	0.67	0.46	0.95	0.47	0.89	0.17	– 0.81	0.71	0.88	0.30	0.05	3.43
Shiga	0.82	0.38	– 0.09	0.63	– 0.86	0.42	0.12	– 1.80	0.35	0.87	0.57	– 0.54	4.45
Kyoto	0.59	0.83	1.17	1.19	1.64	2.14	1.50	– 1.14	1.79	2.00	1.86	0.03	– 1.22
Osaka	– 6.60	1.15	2.46	2.44	3.06	2.23	3.31	0.53	3.41	2.96	3.95	1.97	– 9.98
Hyogo	1.16	0.69	1.20	1.20	1.43	1.61	1.25	2.75	1.29	1.51	2.01	1.66	– 2.32
Nara		0.41	0.62	0.79	0.33	1.13	0.07	– 1.51	0.67	0.42	0.80	– 0.35	5.04
Wakayama	0.43	0.30	0.79	0.70	0.61	1.59	0.53	– 1.80	0.97	1.07	0.79	– 0.40	2.02
Tottori	0.88	0.40	0.34	0.68	0.18	1.32	– 0.09	0.16	0.76	0.71	0.05	– 0.63	3.46
Shimane	0.51	0.27	0.10	0.36	0.13	0.79	– 1.01	0.07	0.21	0.47	0.21	– 0.60	3.48
Okayama	0.58	0.35	0.37	0.84	0.51	0.99	0.31	– 1.43	0.34	0.72	0.75	– 0.37	3.65
Hiroshima	0.69	0.73	0.97	0.75	1.01	1.15	0.05	– 2.70	0.96	0.90	1.30	0.20	0.68
Yamaguchi	0.44	0.40	0.52	0.68	0.51	0.95	0.13	– 1.27	1.01	0.74	0.95	1.24	1.38
Tokushima	1.02	0.05	– 0.22	0.53	0.30	0.73	0.18	– 4.27	0.58	0.76	0.34	– 0.61	3.41
Kagawa		0.38	0.12	0.39	0.43	1.10	– 1.17	– 1.71	0.65	0.91	0.43	– 0.88	3.82
Ehime	– 11.99	0.74	0.62	0.71	0.42	1.31	0.18	– 2.57	0.93	0.82	0.40	– 0.09	3.27
Kochi	0.92	0.58	0.76	0.69	0.70	1.28	0.15	– 1.78	0.49	0.88	– 0.09	– 0.49	2.14
Fukuoka	1.63	0.93	1.70	1.81	1.64	2.03	1.94	4.84	1.02	1.89	1.75	1.99	– 2.02
Saga	1.57	0.80	1.14	0.96	0.41	0.84	– 0.66	1.31	0.32	0.20	– 0.16	0.00	3.90
Nagasaki	1.39	0.91	2.17	2.19	1.65	0.63	1.47	– 1.23	0.48	1.17	1.01	0.67	– 0.34
Kumamoto	1.02	0.70	0.96	0.70	0.56	1.22	– 0.02	– 1.76	1.00	0.88	0.48	– 0.71	3.07

(continued)

Table 5.5 (continued)

Prefecture	1884–88	1888–93	1893–98	1898–03	1903–08	1908–13	1913–18	1918–20	1920–25	1925–30	1930–35	1935–40	1940–45
Oita	0.98	0.29	0.62	0.46	0.50	0.99	– 0.08	– 2.02	1.24	0.66	0.72	– 0.57	3.37
Miyazaki	1.46	1.19	1.14	1.59	1.48	2.07	1.66	1.18	1.20	1.93	1.63	– 0.03	2.11
Kagoshima	1.25	0.96	1.10	1.37	1.38	1.89	0.83	– 0.37	0.79	1.12	0.44	– 0.47	– 0.21
Okinawa	0.63	2.41	1.57	0.72	1.02	1.38	1.52	– 0.10	– 0.49	0.70	0.51	– 0.90	

Statistics Bureau of Japan

in 1920. However, the population of Tokyo city decreased owing to the Great Kanto earthquake in 1923, in which 105,000 people died or missed. On the other hand, Osaka city merged neighboring municipalities in 1925 and became the largest city in Japan. Tokyo city also merged neighboring towns and villages in 1932 and regained the top position. In 1940, four cities in Japan including Nagoya and Kyoto had residents of more than one million, and Kobe and Yokohama were approaching this level. The population size of six cities in Table 5.6 was superior to other local cities. The seventh largest city in 1940 was Hiroshima with 340,000 inhabitants, which was only one-third of Yokohama or Kobe.

Although Zensho (1927) compiled regional populations during the Joseon dynasty, there are no clues as to omission bias by region. Thus, only census populations after 1925 shown in Kim et al. (2002) were used to calculate population growth rates in Table 5.7. While Kim and colleagues separated populations of major cities in today's Republic of Korea according to the administration area in 2000, they did not specify cities in today's DPR Korea. For example, the population of Pyongyang was not separated from that of Pyeongnam and was not used in Fig. 5.2 or Table 5.7.

It is true that the population of major cities in Korea grew faster than the average. The populations of Seoul and Incheon, especially, more than doubled in the late 1930s. However, the Gini coefficient in Fig. 5.2 suggests that the population distribution was equalized during the Japanese occupation. According to Table 5.7, originally densely populated regions such as Gyeongsang province (Gyeongbuk and Gyeongnam) and Jeolla province (Jeonbuk and Jeonnam) showed slower population growth than the average. It seems that a huge number of surplus labor forces migrated out of these regions, resulting in the decline of the Gini coefficient. While some out-migrants headed to cities and sparse northern regions, the industrialization of Korea was too slow to absorb those out-migrants from the southern regions. Thus, a considerable part of them emigrated to Japan, Manchuria, Sakhalin, China, and other countries, as described later.

Japan conducted a full cadastral survey of Korea in 1910–18 and enforced the modern landholding system. During this process, many Koreans were separated from the land to become the surplus population. Relatively expensive fees of farmers' unions also induced the loss of land ownership among peasants. Such problems were especially serious in the southern regions, where rice cropping accounted for most

Table 5.6 Population of major cities in Japan: 1920–40

	1920	1925	1930	1935	1940
Tokyo	2,173,201	1,995,567	2,070,913	5,875,667	6,778,804
Osaka	1,252,983	2,114,804	2,453,573	2,989,874	3,252,340
Nagoya	429,997	768,558	907,404	1,082,816	1,328,084
Kyoto	591,323	679,963	765,142	1,080,593	1,089,726
Kobe	608,644	644,212	787,616	912,179	967,234
Yokohama	422,938	405,888	620,306	704,290	968,091

Census

Table 5.7 Population growth rate (%) by region of Korea: 1925–1944

Region	1925–30	1930–35	1935–40	1940–44
Korea total	1.53	1.69	1.22	1.58
Seoul city	2.10	2.94	16.07	1.39
Busan city	6.50	4.55	6.47	7.15
Daegu city	4.05	2.85	10.74	3.67
Incheon city	3.89	4.02	15.58	5.72
Gwangju city	10.70	6.71	3.39	6.32
Daejeon city	26.22	7.20	3.12	13.91
Gyeonggi	1.07	2.45	− 1.80	1.80
Gangweon	2.23	1.53	1.91	1.28
Chungbuk	1.21	1.28	− 0.31	0.91
Chungnam	1.25	1.88	0.57	1.07
Jeonbuk	1.89	1.34	− 0.11	1.14
Jeonnam	1.56	1.51	0.99	0.87
Gyeongbuk	0.59	1.11	− 1.36	1.11
Gyeongnam	0.76	0.75	− 0.72	1.17
Jeju	0.30	− 0.11	0.64	0.86
Hwanghae	0.83	1.90	1.60	2.65
Pyeongbuk	1.41	1.99	3.77	0.82
Pyeongnam	1.98	1.82	− 0.57	3.14
Hamnam	2.24	1.75	1.76	1.75
Hambuk	3.54	2.74	5.27	0.49

Based on Kim et al. (2002)

economic production (Park 2015). Japan further encouraged the rice monoculture economy in southern parts of Korea by launching the rice production increase plan and abolishing the tariff for Korean rice in the 1920s. Unlike in Taiwan where various crops were available, the rice monoculture in Korea could not absorb the surplus population during the winter (Nakamura 2004). While Koreans who could afford the travel cost headed abroad, others moved into major cities to become manual laborers or small shop-keepers. Seoul experienced over-urbanization accepting more population than the labor demand and those who could not find job concentrated in slums (Kasuya et al. 2016). Even if a migrant could find a job, Korean workers in the secondary industry were more severely exploited than their counterparts in Japan. Korea was assumed to be a paradise for Japanese industrialists, which meant a living hell for Korean workers (Eckert 1991). Thus, the motivation for emigration existed not only in rural but also in urban areas.

As shown in Table 5.8, a city with one million population did not appear in Korea during the Japanese rule. The population of Seoul increased dramatically when neighboring municipalities were merged into the capital in 1936. Nevertheless,

Seoul in 1940 did not match the six major cities of Japan shown in Table 5.6. Chang (2002) wrote that urbanization in this period was delayed by emigration to Japan and other countries. Industrialization was also at an earlier stage than it was in Japan. The percentage of primary industry population was still as high as 74.2% in 1940, compared to 44.3% in Japan. According to Cumings (2005), Korea was still fundamentally an agrarian society in 1945.

It was after the annexation to Japan that the whole island of Taiwan was integrated politically and economically. Although Taiwan was the first colony of the Japanese empire, the colonization of Taiwan was more successful than that of Korea. Japan established land and population registers, banks and a financial system, various communication and transportation systems, and an administrative system based on police stations and paojias. Based on this infrastructure, the agricultural system of Taiwan developed dramatically to produce various crops.

Before the annexation to Japan, the east coast of Taiwan was a sparsely inhabited area mainly dwelled by indigenous people. Equalization of the population distribution during the Japanese era occurred through the colonizing migration to eastern prefectures such as Karenko and Taitung, as suggested in Table 5.9. Although the high population growth rates of these two provinces caused the decline in Gini coefficients in Fig. 5.2, more than 95% of the population in 1940 still inhabited the western parts.

As shown in Table 5.10, the growth of major cities in Taiwan was very slow. Even the capital of Taipei held less than half of the population of the Korean capital of Seoul in 1940. Unlike in Japan and Korea where a huge outflow of surplus population from rural areas took place, the Taiwanese had little motivation to out-migrate from their hometowns because the agricultural development was very successful. The Governor-General encouraged the planting of various crops, including sugarcane, tobacco, tea, bananas, and pineapples, in addition to rice. These agricultural products were processed and exported to Japan and other nations. Thanks to such competitive agriculture, the growing population in Taiwan could find job opportunities without out-migrating from villages. While peasants in Japan and Korea confronted a difficult situation, capital accumulation proceeded, and wealth distribution was equalized in rural Taiwan.

Table 5.8 Population of major cities in Korea: 1925–44

	1925	1930	1935	1940	1944
Seoul	247,404	279,865	312,587	775,162	824,976
Pyongyang	70,075	116,899	154,759	254,599	306,685
Busan	64,653	97,558	130,017	192,215	267,187
Incheon	41,541	52,971	67,126	151,454	190,669
Daegu	58,411	73,060	85,453	158,468	185,465
Cheongjin	13,471	25,639	42,145	165,869	151,740

Kwon et al. (1975)

Table 5.9 Population growth rate (%) by region of Taiwan: 1925–1940

Region	1925–30	1930–35	1935–40
Taiwan total	2.84	2.56	3.12
Taipei	2.31	2.32	3.09
Hsinchu	2.15	1.39	2.75
Taichung	3.24	2.74	3.08
Tainan	2.41	2.81	2.74
Kaohsiung	1.63	2.96	3.94
Karenko	8.07	5.12	5.66
Taitung	6.62	3.76	4.58
Penghu	–	1.92	0.50

Census

Table 5.10 Population of major cities in Taiwan: 1925–44

	1925	1930	1935	1940
Taipei	195,200	230,490	274,157	353,744
Kaohsiung	43,764	62,722	85,467	161,418
Tainan	84,793	94,546	110,816	149,969
Keelung	62,000	75,070	86,887	105,084
Taichung	41,042	54,188	70,069	87,119

Census

Table 5.11 shows the percentage of urban population calculated from the census results in the Empire of Japan. Although the major cities in Korea had larger population than those in Taiwan, the percentage of urban population was smaller. The percentage increased by almost 20% points in Japan between 1920 and 1940, which significantly raised the Gini coefficient. While the population of five cities in Taiwan in 1920 accounted for 10.7% of the total population, the sum of twelve cities in Korea accounted for only 5.2%. By 1940, the number of cities increased to eleven in Taiwan and twenty in Korea. However, the increase in the percentage of urban population was overridden by the outflow from densely inhabited rural areas and had no impact on the Gini coefficient.

Table 5.11 Percentage of urban population: 1920–40

	Japan	Korea	Taiwan
1920	18.0		
1925	21.6	5.2	10.7
1930	24.0	6.3	13.5
1935	32.7	7.0	16.2
1940	37.7	11.6	19.2

Census

The 1930 census of the Empire of Japan contained contingency tables of place of birth and the current residence. Table 5.12 compares the percentage living in the region of birth. Because the denominator is the place of birth, a lower percentage implies a higher out-migration rate. The percentage is highest for the Taiwanese; more than 95% were living in the place of birth. While the number of overseas Taiwanese was negligible, some adjustments for the overseas population would be required for Korea and Japan. According to the same census, 419,009 Koreans and only 4,611 Taiwanese lived in Japan. It is also assumed that slightly more than 600,000 Koreans lived in Manchuria in 1930 (Kim 1965; Park 2009). If we add one million overseas Koreans to the denominator, the percentage will drop from 92.8% in Table 5.11 to 88.4%. According to Wakatsuki (1995), there were 1,775,896 overseas Japanese in 1930. If this amount were added to the denominator, the percentage living in the place of birth would be 81.3% rather than 83.6% in the table. Thus, it is apparent that the Taiwanese had the significantly lower propensity to move than the Japanese or Koreans.

As mentioned above, changes in the Gini coefficient of China in Fig. 5.2 do not reflect the urbanization. The decline in 1845–65 could be attributed to the fall in densely inhabited provinces such as Anhui, Jiangsu, Zhejiang, and Fujian, as suggested in Table 5.13. These provinces were the main battlefields in the Taiping Rebellion and showed higher population growth rates than the national average in 1875–85, suggesting a massive return migration flow. The decline in the Gini coefficient in 1912–28 was caused by population decreases in Fujian, Jiangxi, Hubei, Shandong, and Zhejiang. The Qing dynasty finally fell in 1912, and the Republic of China was founded. However, the central government could not control the nation, and many warlords took power in each local area after the death of Yuan Shikai in 1916. While such political turmoil induced the equalization of population distribution, the Chinese economy started growing rapidly after WWI. The textile industry and other light industries in coastal provinces led the economic growth, encouraging population growth in Jiangsu, Fujian, and Shandong. As a result, the Gini coefficient rose again in 1928–36.

According to Table 5.14, the population of Shanghai was overwhelmingly larger than other major cities in the early twentieth-century China. Beijing was renamed "Beiping" by the Republic of China in 1928 because the city was not the capital (jing) anymore. Instead, Xi'an was defined as the second capital of the Republic and called "Xijing" in 1932–43. The economic and demographic superiority of Shanghai was formed by the fact that the city was the gateway of invasion by the Western powers. Shanghai was opened as one of five treaty ports defined in the Treaty of Nanjing (1842). When the Taiping rebels occupied Suzhou in 1860, Shanghai replaced its position to become the economic center of the Changjiang basin. The machine manufacturing industry in China started from Shanghai and silk mills, weaving mills, and spinning factories were built in the late nineteenth century (Okamoto 2013). At the beginning of the Republic of China, the economic basis of the government was fragile. After WWI, however, the Chinese economy grew steadily led by the light manufacturing, financing, and real estate business in Shanghai (Kajitani and Kashima 2013).

Table 5.12 Percentage living in the place of birth: 1930

Japan				Korea		Taiwan	
Prefecture	(%)	Prefecture	(%)	Region	(%)	Region	(%)
Total	83.6			Total	92.8	Total	95.7
Hokkaido	92.9			Gyeonggi	91.3	Taipei	96.3
Aomori	84.6	Shiga	74.2	Chungbuk	85.6	Hsinchu	87.5
Iwate	86.8	Kyoto	84.2	Chungnam	91.4	Taichung	98.3
Miyagi	82.0	Osaka	87.6	Jeonbuk	94.0	Tainan	98.2
Akita	81.8	Hyogo	83.9	Jeonnam	96.1	Kaohsiung	98.3
Yamagata	84.7	Nara	73.2	Gyeongbuk	91.5	Karenko	95.7
Fukushima	83.0	Wakayama	82.0	Gyeongnam	94.7	Taitung	96.5
Ibaraki	82.0	Tottori	82.4	Hwanghae	93.1	Penghu	77.8
Tochigi	79.1	Shimane	84.4	Pyeongnam	88.8		
Gunma	83.7	Okayama	83.6	Pyeongbuk	94.8		
Saitama	80.1	Hiroshima	81.5	Gangweon	92.4		
Chiba	80.7	Yamaguchi	84.7	Hamnam	93.5		
Tokyo	86.7	Tokushima	78.2	Hambuk	96.3		
Kanagawa	82.7	Kagawa	75.4				
Niigata	79.3	Ehime	81.3				
Toyama	73.4	Kochi	85.8				
Ishikawa	75.1	Fukuoka	89.6				
Fukui	75.5	Saga	77.0				
Yamanashi	79.1	Nagasaki	87.2				
Nagano	87.7	Kumamoto	85.2				
Gifu	79.7	Oita	83.1				
Shizuoka	87.4	Miyazaki	90.2				
Aichi	89.5	Kagoshima	87.0				
Mie	81.5	Okinawa	92.9				

The 1930 census

5.7 Cross-Border Migration in Modern Eastern Asia

While the invasion of barbarians into northern China continued from the ancient era, the mass outflow from southern China to Southeastern Asia became prominent in the medieval period. The overseas Chinese population in Southeastern Asia was already noticeable in the sixteenth century. After the Qing dynasty was opened to the West, a large number of Chinese emigrants flowed into the US and Australia during the gold rush. When the US, Canada, and Australia restricted the Chinese migrants at the end of the nineteenth century, Japanese migrants increased instead, until they were also restricted in the early twentieth century. The expansion of the Empire of

Table 5.13 Population growth rate (%) by province of China: 1845–1936

Province	1845–55	1855–65	1865–75	1875–85		1912–28	1928–36
China total	0.06	– 1.52	– 2.34	0.31		0.90	– 0.09
Hebei	0.05	0.20	1.26	0.38		0.99	0.07
Shanxi	0.64	0.22	0.40	– 4.10		1.21	– 0.66
Rehe						2.23	– 12.89
Chahar						1.31	0.24
Suiyuan						7.89	– 0.24
Inner Mongolia	0.32	0.32	3.22	0.32		19.85	0.00
Liaoning	1.07	0.39	21.78	2.24		1.43	0.02
Jilin	0.03	0.12	2.11	2.45		1.98	– 0.47
Heilongjiang	0.30	0.38	5.11	4.61		3.87	2.19
Jiangsu	– 0.77	– 5.22	– 15.32	1.14		0.35	2.39
Zhejiang	0.36	– 10.71	15.82	1.73		– 0.24	0.35
Anhui	– 1.58	– 6.76	– 14.71	0.50		1.84	0.91
Fujian	0.56	– 0.58	8.73	1.29		– 2.79	1.95
Jiangxi	– 0.20	0.19	0.09	0.01		– 1.03	– 3.09
Shandong	0.36	0.25	2.64	0.30		– 0.48	3.87
Henan	0.07	0.00	0.03	0.59		0.43	1.45
Hubei	– 0.45	– 0.05	2.97	0.26		– 0.64	– 0.56
Hunan	0.19	0.12	0.02	0.00		0.83	– 1.33
Guangdong	0.70	0.09	0.95	0.06		0.92	0.01
Guangxi	0.05	– 0.12	– 0.99	– 0.10		3.49	– 0.24
Sichuan	1.34	– 0.52	4.62	0.10		1.05	1.18
Xikang							– 24.23
Guizhou	– 1.07	– 1.22	9.67	2.01		2.68	– 4.84
Yunnan	0.41	– 2.91	– 14.10	1.85		2.39	– 1.71
Tibet	0.14	0.14	1.44	0.14		7.56	0.00
Shaangxi	0.02	– 1.40	– 25.65	0.61		1.46	– 2.07
Gansu	0.02	– 1.68	– 64.23	0.01		1.45	0.84
Qinghai	0.18	– 0.19	– 4.90	0.56		19.30	– 18.58
Ningxia	0.02	– 1.86	– 77.01	– 0.48		10.28	– 4.80
Xinjiang	1.74	1.81	18.58	1.72		1.23	6.92

Based on Zhao and Xie (1988) cited in Ueda (1995), Minami and Makino (2014)

Japan had an enormous impact on cross-border migration in Eastern Asia. While the high population pressure resulted in substantial flows from Japan and Korea, the Taiwanese villagers could stay thanks to competitive agricultural production. Because the territory of a nation frequently changed during the age of imperialism, the term "cross-border" is preferred here to "international" migration.

Table 5.14 Population of major cities in China (thousand)

	1936–37	1947
Shanghai	3,727	4,630
Beiping	1,551	1,722
Tianjin	1,218	1,773
Nanjing	1,019	1,114
Qingdao	515	851
Xian (Xijing)	206	828

Minami and Makino (2014)

IPSS (2023) contains a table of population growth and natural growth numbers for the Japanese population in 1872–1920, citing the Statistics Bureau of the Cabinet Office (1930). For years after 1920, IPSS compiled these numbers for total population. These data were used to calculate the net immigration rate shown in Fig. 5.3. While the net immigration rate for the Japanese was close to zero until 1895, significant emigration was observed when Japan annexed Taiwan (1895) and obtained the superior power in Korea (1905). Net immigration for the total population was close to zero during the 1920s, which implies that emigration of Japanese was offset by immigration of non-Japanese including Koreans. However, emigration accelerated in the 1930s, and the rate exceeded − 2 per thousand after 1935. Military expeditions in the Second Sino-Japanese War and Pacific War apparently accelerated emigration.

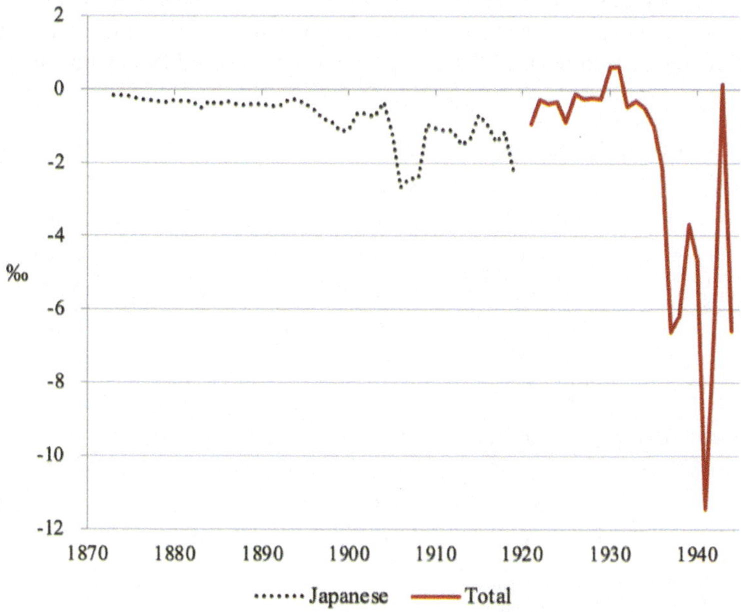

Fig. 5.3 Net immigration rate of Japan: 1873–1944. IPSS (2023)

Shiode (2015) compiled the number of overseas Japanese reported by the Ministry of Foreign Affairs. Wakatsuki (1995) cited the data of Japanese repatriation, edited by the former Ministry of Health and Welfare. These data are shown in Table 5.15. The number of overseas Japanese in 1940 amounted 3.323 million, or 4.6% of the total population in Japan. According to the repatriation data, the number increased to 6.295 million by 1945. The count could be higher if we consider those who did not report after landing Japan or did not return at all. Many Japanese soldiers and civilians were captured when the Soviet Union invaded Manchuria in August 1945. They were interned in Siberia and approximately 470,000 survived and returned to Japan.

The largest number of Japanese lived in Manchuria in 1940, and the largest number of civilians repatriated from there. Japan conquered Manchuria and founded Manchukuo, a puppet state, in 1932. Manchukuo was assumed to be a colony of Japan, and many administrators and experts were sent from Japan. The high economic growth rate attracted Japanese capitalists, merchants, and laborers. The Japanese government recruited pioneers to cultivate remote wastelands, and 42,636 households of farmers and 97,030 young volunteer soldiers were sent in 1937–41 (Kita 1944).

After the First Sino-Japanese War, Japan built concessions in Tianjin, Hankou, Hangzhou, Suzhou, and Chongqing in addition to the international settlement in Shanghai shared with the UK and the US. Japan sent the largest number of soldiers to the Boxer Rebellion and succeeded the German concession in Shandong province by enforcing the Twenty-One Demands (1915). The Japanese population in China increased under these processes. Although the number of Japanese residents in mainland China was smaller than those in Korea or Manchuria in 1940, the largest number of soldiers repatriated from China. More civilians repatriated from China than from

Table 5.15 Overseas Japanese and repatriation

Region	Oversea Japanese	Repatriation after 1945		
	(1940)	Total	Military	Civilian
Soviet Union[a]	381,614	766,441	469,793	296,648
Manchuria[b]	1,065,072	1,271,479	52,833	1,218,646
China[c]	284,680	1,559,238	1,058,745	500,493
Korea	689,747	919,904	206,600	713,304
Taiwan	346,663	479,544	157,388	322,156
Southeastern Asia	36,467	892,526	807,081	85,445
Pacific Islands	81,011	130,968	103,462	27,506
Other	437,884	275,104	251,509	23,595
Total	3,323,138	6,295,204	31,07,411	31,87,793

[a] Includes Sakhalin and Kuril islands
[b] Includes Kwangtung Leased Territory
[c] Includes Hong Kong
Shiode (2015), Wakatsuki (1995)

Taiwan. It is supposed that various government-controlled companies in investment, development, and transportation founded during the Second Sino-Japanese War attracted a large number of Japanese workers (Wakatsuki 1995).

There were more overseas Japanese in Korea and Taiwan in 1940 than in mainland China. Since these colonies were occupied for decades, the increase in Japanese residents in 1940–45 was smaller than in other newly conquered areas. Unlike in the Asian colonies of Western powers, Japanese migrants dwelled not only in urban but also in rural areas. Even in small towns in Korea, there lived five or six Japanese, including a policeman, headmaster and teachers at the primary school, and officials of the water union and financial associations (Lee 2009). According to Barclay (1954), the percentage of Japanese in Taiwan living in the nine major cities decreased from 20.6% in 1920 to 19.2% in 1940. By contrast, the population of Westerners in Asian colonies was small and concentrated in cities. According to Cumings (2005), French residents in Vietnam accounted for only 0.8% of total population in 1937 and the percentage was even smaller in British colonies.

After the Pacific War erupted in December 1941, Japan invaded British Malaya, the US Philippines, and the Dutch East Indies. Singapore was occupied in February 1942. By the end of 1942, the occupied territory of Japan expanded to eastern New Guinea, New Britain, Midway, the Fiji Islands, and Samoa. During this phase of war, Japanese inhabitants of Southeastern Asia and the Pacific islands increased dramatically. As in Table 5.15, while there were 117,478 Japanese in these areas in 1940, more than one million soldiers and civilians repatriated to Japan after the war.

It was assumed that bad cropping and serial rebellions in the nineteenth century stimulated emigration from northern Korea to Manchuria (Kwon 2002). The Qing dynasty gave up border control in the late nineteenth century and permitted Korean peasants to cultivate wastelands (Park 2009). In 1910, there were more than 200,000 Koreans living in Manchuria (Kim 1965). The government of the Republic of China encouraged migrants from northern China to Manchuria. This policy also attracted Korean migrants. After the March 1st Movement in 1919, many Korean independence activists settled in Manchuria and carried on the resistance against Japan. Despite strengthened restrictions against Korean migrants, emigration to Manchuria continued owing to the high population pressure (Park 2015). When Manchukuo was founded in 1932, emigration accelerated because of relaxed restrictions and the encouragement of frontier cultivation. As a result, the Korean population in Manchuria surpassed one million by 1940. Park (2009) estimated that there were 1,948,375 Koreans in Manchuria in 1945.

As shown in Table 5.16, the number of Koreans living in Japan increased drastically from 41,000 in 1920 to 1.3 million in 1940, which accounted for more than 95% of non-Japanese population in Japan. The government attempted various measures to control such a massive flow from Korea. In 1919, the Governor-General of Korea ordered Koreans who planned to travel out of Korea to apply for permission at police stations. This procedure was abolished in 1922. Instead, Koreans without a workplace in Japan, a defined amount of money, or Japanese fluency were prevented from boarding at Busan as of 1925. Since 1928, Koreans were required to apply for certification to migrate to Japan (Mizuno and Moon 2015). Despite these restrictions, the number of Koreans in Japan increased constantly. As of 1939, Korean laborers were mobilized to Japan to cope with the labor shortage due to the war. As a result, the Korean population at the end of the war was estimated to be 2.1 million (Park 1957). Koreans settled also in other places than Manchuria and Japan. Although the detailed size is unknown, there were major Korean communities in Sakhalin, Primorskaya Oblast, Kazakhstan, and Uzbekistan. Korean communities in mainland China, Europe, and Northern America must have been smaller than them.

While internal mobility of the Taiwanese was low, Taiwan was also virtually closed to cross-border migration. Japan isolated Taiwan from mainland China and controlled the border strictly. Thus, the ethnic composition of the 1940 census was 93.5% Taiwanese, 5.7% Japanese, and only 0.8% Chinese with the nationality of the Republic of China. As Table 5.16 shows, there were only 22,500 Taiwanese in Japan compared to 1.2 million Koreans in 1940. While there was some development in the light industry in Taiwan during the 1930s, capitalists had difficulty to recruit workers because villagers did not want to out-migrate from rural areas. Although the business sector tried to invite Chinese works from Fujian province, the Governor-General's office strictly restricted immigrants to maintain the public order. As a result, the number of invited Chinese workers was too small to satisfy the labor demand (Oshima 2015). Although Japan mobilized some Taiwanese laborers during the Pacific War, the Taiwanese population in Japan at the end of the war would have been around 35,000 (Cumings 1997). The differential mobility between Koreans and Taiwanese evolved into the different modes of urbanization in the Republic of Korea and the Republic of China after WWII.

Until the early twentieth century, Manchuria functioned as the frontier accepting migrants from northern China. Although the Qing dynasty banned Han Chinese to settle in Manchuria, illegal migrants from Zhili (Hebei) and Shandong provinces increased as of the Kangxi period (1661–1722). An attempt to send Manchurian settlers did not work, and the palace had to permit Han migrants to defend Manchuria against Russia (Komine 2014). In fact, the migration from northern China to Manchuria was larger than the emigration from southern China to Southeastern Asia. While there were more than 6.0 million overseas Chinese in 1926 (Chen 2013), most of the 25.5 million Manchurian population in 1925 was the offspring of migrants from northern China (Takaoka and Uehara 1943).

Table 5.16 Non-Japanese population in Japan by ethnicity or nationality

	1920	(%)	1930	(%)	1940	(%)
Non-Japanese total	78,061	(100.0)	477,980	(100.0)	1,304,286	(100.0)
Koreans	40,755	(52.2)	419,009	(87.7)	1,241,315	(95.2)
Taiwanese	1,703	(2.2)	4611	(1.0)	22,499	(1.7)
Sakhaliens	31	(0.0)	22	(0.0)	986	(0.1)
Pacific islanders	3	(0.0)	18	(0.0)	249	(0.0)
Foreigners	35,569	(45.6)	54,320	(11.4)	39,237	(3.0)

Census

As already shown in Table 5.13, the Manchurian provinces of Liaoning, Jilin, and Heilongjiang always sustained higher population growth rates than the average until 1928. The low population growth rates of Liaoning and Jilin in 1928–36 would be erroneous. While Minami and Makino (2014) relied on statistics of the Nationalist government for the population in 1936, the government had no influence in Manchukuo. As seen in Table 5.17, Japanese sources indicate that the Manchurian population grew at a rate of around 2% by 1930, and growth accelerated after Manchukuo was founded. The extremely high growth of 4% or more implies that there was a major flow from northern China. The total population of Manchuria grew from 30.7 million in 1932 to 46.6 million in 1942. According to Yamanaka's estimation, 86.8% of the increase was attributed to the Chinese (including Manchurian, Mongolian, Hui, and other minority groups), 7.1% to Koreans, and 3.5% to the Japanese.

The Chinese migration to Southeastern Asia started during the Song dynasty. By the end of the sixteenth century, there were major Chinese communities with more than 10,000 inhabitants in Luzon and Java. The huge migration during the gold rush mainly came from Guangdong province. The migration to Southeastern Asia increased drastically when the emigration was officially legalized in the Convention of Peking in 1860 (Chen 2013). The National Overseas Affairs Commission (NOAC) reported in 1934 that there were 7.8 million overseas Chinese worldwide (Sugihara 1994). This was an increase of 30% from 6.0 million in 1926. Apparently, the number included those who did not have the nationality of the Republic of China. For example, a child born in Thailand was assigned the Thai nationality and could not have the nationality of a foreign parent. Thus, while the NOAC reported that there were 2.5

Table 5.17 Population estimates for Manchuria

Takaoka and Uehara (1943)			Kondo (1942)			Yamanaka (2005)		
Year	Population (1000)	Annual growth rate (%)	Year	Population (1000)	Annual growth rate (%)	Year	Population (1000)	Annual growth rate (%)
1903	15,290	2.3	1907	16,778	2.3			
1908	17,156	2.3	1912	18,774	2.3			
1913	19,208	2.3	1917	21,069	2.4			
1918	21,569	2.4	1921	23,156	2.4			
1923	24,295	2.9	1926	26,133	2.8			
1928	28,034	1.1	1930	29,198				
1933	29,606					1932	30,655	4.9
						1935	35,386	4.1
						1940	43,203	3.8
						1942	46,564	

Notes Takaoka and Uehara relied on statistics by Minamimanshu Tetsudo for 1903–28, and on Asia Nenkan for 1933 Kondo cited Ishida (1932) Yamanaka estimated detailed population structure based on official population statistics million Chinese in Thailand, the number of first-generation Chinese sustaining the Chinese nationality was only 445,000 (Kikakuin 1939).

While the absolute number of oversea Chinese was large, its ratio to the homeland population was smaller than the case of oversea Japanese or Koreans. During the modern area, Korea experienced the largest diaspora in Eastern Asia. In 1940, there were 1.24 million Koreans in Japan (see Table 5.16) and 1.45 million in China and Manchukuo (Kwon 2002). If we assume that the total number of overseas Koreans was approximately 2.8 million, this represents 11.5% of the population in Korea. The ratio of overseas Japanese in 1940 (3.3 million) to the census population in Japan (71.8 million) was 4.6%. The ratio declined radically because of repatriation after the war. Although the number of overseas Chinese, 7.8 million in 1934, was huge, the domestic population of China was so large that the ratio of overseas population did not exceed that of Japan. Dividing 7.8 million with the population of 532.2 million in 1935 (see Table 5.4) gives 1.47%. It seems apparent that the Taiwanese were the least mobile people during the modern era. Even if we assume that the total number of overseas Taiwanese in 1940 was twice as much as that in Japan, the ratio was less than 1%.

5.8 Demographic Transition

Former sections discussed population growth and distribution, referring to censuses and other statistics on population size by country or region. To discuss the demographic transition, however, vital statistics on births and deaths are required. Unfortunately, reliable vital statistics in the modern period were available only in Japan and Taiwan. The coverage of birth and death certificates in Korea was too low to know the actual level of fertility and mortality. As in the total population, estimated vital rates by Korean demographers varied in a wide range. There were no vital statistics either in mainland China or Manchukuo. The only clue for vital rates in modern China is the indirect estimates by Barclay and his colleagues (1976).

5.8.1 Crude Birth Rate and Crude Death Rate

Figure 5.4 shows the trend of CBR and CDR in Japan and Taiwan. Although CBRs and CDRs by the Statistics Bureau of Japan show long-term increases in birth and death rates in the 1870s and 1880s, a considerable part should be attributed to the improvement of coverage. Japanese demographers attempted to remove the effect to obtain the genuine trends of fertility and mortality. In a typical process of demographic transition, such as that in Sweden, mortality declined first while fertility remained at a high level, which resulted in the modern population increase. This pattern was observed Taiwan after the annexation to Japan. Although there were some temporary rises due to the plague (1914–16) and the influenza (1918–20), the

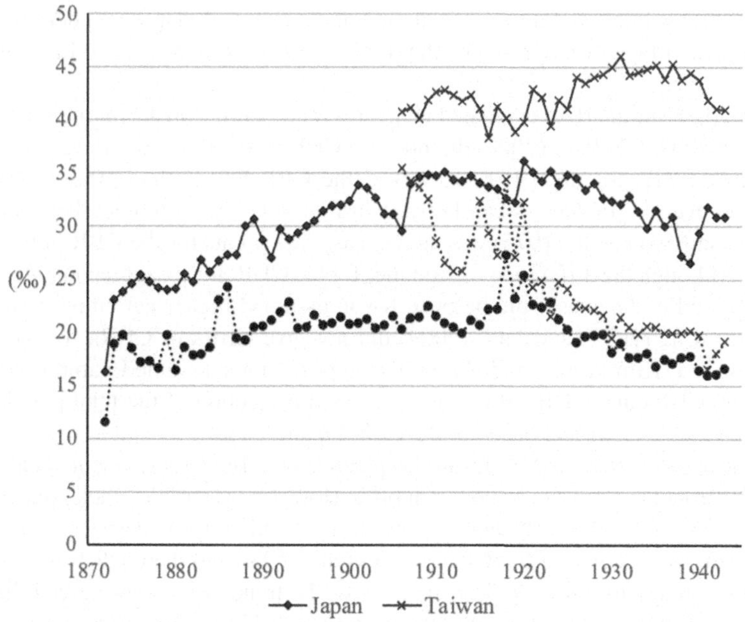

Fig. 5.4 Crude birth and death rates: 1872–1944. Statistics Bureau (1930), Mizoguchi (2008). Solid line = CBR, Dotted line = CDR

CDR continued to decline during the Japanese era. On the other hand, the CBR stayed at a high level and the population growth accelerated. In Japan, however, the CDR fluctuated within the range between 20 and 23 per thousand in 1890–1917 and rose sharply owing to the influenza pandemic in 1918–20. It was after the mid-1920s that the continuous decline of mortality was observed. Thus, as far as we rely on the official vital statistics, the demographic transition was not led by the mortality decline but the rise in fertility.

Some Japanese demographers including Yasukawa (1977) and Okazaki (1986) did not believe the vital statistics before the 1920 census but presumed that the orthodox demographic transition process led by the decline in death rate took place in Japan. They assumed that the CDR started declining at least by the 1880s and constructed life tables in the past according to this assumption. Takase (1991) criticized those backward projections assuming the long-term mortality decline. According to him, projections by Yasukawa and Okazaki showed unnatural age patterns of differences from the registered population. Furthermore, their projections implied that the coverage of birth and death certifications was prominently improved not only in the 1880s but also in 1905–15. While the improvement in the 1880s can be explained by the proclamation of a new burial rule in 1884, there is no reason to support the second improvement because the coverage was already high in the early twentieth century.

Saito (2001) supported Takase's discussion and concluded that the modern population growth in Japan was caused by the rise in fertility, as in the case of England and Wales.

Table 5.18 summarizes estimated vital rates for modern Korea. Among six studies shown in Table 5.2, Park (2008) did not give CBRs or CDRs. Ishi (1972) estimated very high CBRs for the 1910s. However, the CBR higher than 50‰ was never observed even in Taiwan as can be seen in Fig. 5.4. The difference between Ishi (1972) and Kwon et al. (1975) was as large as 15‰ points for the CBR in the early 1910s. Although the difference was smaller for the CDR, the disagreement remained until the end of Japanese rule because Kwon assumed higher net emigration rates than Ishi. Cha (2009) and Park (2009) did not give CBRs or CDRs for the early Japanese era. Kim (1965) and Kwon discomposed their assumed natural increase rates into CBRs and CDRs after their backward projection of the total population. They gave little empirical basis for the early Japanese era.

Because Ishi conducted backward projections with life tables, he could calculate the CBR from the size of each cohort at birth. However, as Park Yi-Taek pointed out, Ishi's estimates for the early Japanese era were too high. Considering the situation of Taiwan, it is likely that the CBR was less than 50‰ and the CDR was less than 35‰ in Korea before the 1925 census. As for the trend, all studies agreed that the death rate declined while the birth rate remained at a high level during the Japanese era. Thus, it is safe to say that Taiwan and Korea experienced the first stage of the orthodox demographic transition in the first half of the twentieth century.

Table 5.18 Estimations for Korean vital rates in Japanese era

	Official	Ishi (1972)	Kim (1965)	Cha (2014)	Park (2009)	Kwon et al. (1975)
Crude birth rate (‰)						
1910–15	26.34	53.02	42			38
1915–20	31.49	52.09	42			40
1920–25	36.32	48.82	42.8		43.04	42
1925–30	37.52	48.53	42.6	44.9	42.25	45
1930–35	30.93	47.89	42.4	44.1	42.94	44
1935–40	32.57	46.97	42.3	43.8	42.44	44
1940–45	35.38	42.79	41.6	42.1	43.12	42
Crude death rate (‰)						
1910–15	17.24	37.61	32			34
1915–20	24.73	34.36	34.6			33
1920–25	20.82	32.19	25.7		28.63	30
1925–30	21.58	30.28	25.0	26.2	24.80	26
1930–35	20.29	26.45	21.6	23.9	22.31	24
1935–40	18.37	21.28	21.4	23.2	21.99	23
1940–45	18.50	19.32	18.9	21.9	18.54	23

Vital rates of Taiwan by Mizoguchi (2008) shown in Fig. 5.4 basically accepted the official vital statistics, with a slight upward revision due to an adjustment for neonatal mortality in 1906–16. The level of mortality before the annexation to Japan must have been much higher than it was in 1906, because Taiwan was notoriously unhealthy island. After the crisis of the plague and the influenza pandemics, the CDR of Taiwan dropped to a level close to that of Japan. Figure 5.4 shows the rise in fertility in the 1920s in Taiwan. The Gross Reproduction Rate (GRR) of Taiwan estimated by Barclay (1954) rose from 2.93 in 1905 to 3.39 in 1930. It is likely that the rise in fertility was caused by the rural development owing to a variety of commercial crops. Because Koreans were less happy under the Japanese rule than the Taiwanese, the CBR in Korea during the 1930s was less than that in Taiwan, as suggested by Kim Cheol and Park Keong-Suk.

Unlike in the Empire of Japan, there was no vital statistics in the Republic of China. The only reliable estimation of vital rates is an indirect estimation by Barclay and his colleagues (1976). The estimation was based on a large size sample survey conducted by the University of Nanjing in 1929–31. The data included the dates of marriages, childbirths, and deaths of household members in 119 rural villages in 16 provinces other than Manchuria and southern coastal provinces. The estimated CBR and CDR were 41.2‰ and 41.5‰, respectively. The result implies that the Chinese population stagnated in this period, and that mortality was considerably higher than in the domain of the Japanese empire.

5.8.2 *Nuptiality, Fertility, and Mortality*

Table 5.19 summarizes indicators of nuptiality, fertility and mortality around 1930 when the estimates for China are available. The Singulate Mean Age at Marriage (SMAM) was proposed by Hajnal (1953) as an estimate of the mean age at first marriage based on the proportion of never-married men or women in a census. According to Saito (1997), a female SMAM higher than 20 was only observed in Japan and Sri Lanka (Ceylon, at that time) in the early twentieth-century Asia. As can be seen in the Table, Korean women married especially early. Kim (2008) pointed out that women in teens accounted for three-quarters of annual marriages in Korea during the Japanese era, which was higher than any country other than India. Kim explained that an early marriage was guaranteed in the Joseon dynasty because consanguineous marriages and intermarriages between social classes, political factions, and birth statuses (legitimate or illegitimate) were strictly limited. The Gabo Reform in 1894 defined the minimum marriage age at 20 for men and 16 for men to prohibit child marriages. However, the law was ignored, and marriages of low teens remained dominant (Zensho 1927).

The Taiwanese women married considerably later than Korean women in the Japanese regime. The Rinji Taiwan Kyukan Chosakai (Temporary Committee on Old Customs in Taiwan) reported in 1903 that it was unusual that men married under 18 and women married under 16 years of age. Thus, it is likely that the female mean

Table 5.19 Demographic indicators around 1930

	Japan	Korea	Taiwan	China
Singulate mean age at marriage (male)	25.8[a]	20.8[a]	23.6[a]	21.3[h]
Singulate mean age at marriage (female)	21.8[a]	16.5[a]	19.4[a]	17.5[h]
Total fertility rate	4.7[b]	6.4[c]		5.5[h]
Infant mortality rate (‰)	118.2[b]	151[d]	148.3[f]	296[h]
Life expectancy at birth (male)	44.8[b]	32.4[d]	38.8[g]	24.6[h]
		37.9[e]		
Life expectancy at birth (female)	46.5[b]	34.9[d]	43.1[g]	23.7[h]
		37.2[e]		

[a] Author's calculation based on the 1930 census
[b] IPSS (2023)
[c] Park (2009)
[d] Ishi (1972)
[e] Kwon et al. (1975)
[f] Mizoguchi (2008)
[g] Barclay (1954)
[h] Barclay et al. (1976)

age at marriage was already higher than in Korea at the beginning of the twentieth century. According to Barclay (1954), the proportion of never-married women under age 35 constantly increased in Taiwan between 1905 and 1935. On the other hand, the mean age at marriage of men declined as the unbalanced sex ratio in marriageable ages was gradually resolved. One possible reason for the reduced marriage squeeze would be the decrease in female infanticide under the Japanese rule.

The female mean age at marriage in rural China around 1930 was within the range of Daoyi village in the nineteenth century. The delay in marriage proceeded even in Korea, and the SMAM rose to 17.8 in 1940 (Kwon and Kim 2002). In rural China, however, it is doubtful that such modern changes as mortality decline, the decrease in child marriage, the rise in school enrollment rate, and the female labor participation in the secondary and tertiary industries was progressing.

While many European countries experienced below-replacement fertility during the interwar period, the TFR of Eastern Asia remained at a high level. The replacement level is the TFR that eventually result in the zero-population growth, corresponding to the Net Reproduction Rate of one and the intrinsic natural increase rate of zero. The TFR of Japan in 1930 was 4.7, which was 1.5 times higher than the replacement level and implied that the population would grow by 50% in one generation. Although Japan was the front runner of the demographic transition in Asia, fertility decline was so slow that the TFR remained at 4.11 in 1940. Park (2009) estimated the TFR with 35 * General Fertility Rate (GFR) = 35 * (number of births)/(female population aged 15–49 years old). The assumption is that the GFR is a good estimator of the average of age-specific birth rates. Although no estimate of TFR was found for Taiwan under the Japanese rule, Barclay (1954) estimated that the GRR was 3.39 in 1930. The estimated TFR would be 3.39 * 2.06 = 6.98, if we assumed that the sex ratio at birth was 1.06. It is likely that the NRR was close to two in Korea and Taiwan, implying that the population would be doubled in one generation. On the other hand, the TFR of 5.5 in rural China was not sufficient to avoid the long-term population decline. Barclay and colleagues concluded that the intrinsic rate of natural increase was negative in 1930.

According to Maddison (2001), the life expectancy at birth in Japan was 44 years in 1900, which was slightly shorter than the average of Western Europe (47 years). Thus, it is possible that Japan's mortality in the interwar period was lowest in Asia and Africa. Korea and Taiwan within the Japanese Empire could enjoy the fruit of modernization such as the introduction of western medicine, the infrastructure for public health, and the spread of hygienic habits through education. However, the impact of modernization was considerably smaller in rural China. For mortality in Korea, controversy remained even after the modern census in 1925, and the difference was wide between Ishi who assumed mostly closed population and Kwon who assumed a high net-emigration rate. Interestingly, however, the wide range of estimates given by these two studies stayed within the range between Taiwan and China. Higher mortality in Korea than in Taiwan could be explained by the economic condition. It is likely that the nutritional condition was improved faster in Taiwan where the agricultural economy was better than in Korea.

5.8.3 Causes of Deaths

In his classic article on the epidemiological transition, Omran (1971) discussed that mortality decline passed three stages: (1) Age of Pestilence and Famine, (2) Age of Receding Pandemics, and (3) Age of Degenerative and Man-Made Diseases. Three stages basically correspond to the premodern period, the modern period, and the contemporary period, respectively. Omran supposed that the second stage was characterized by the modern population increase caused by the decrease in deaths from infectious and epidemic diseases. However, mortality did not always decline, and infectious diseases did not lose the influence in modern Eastern Asia. In Japan, mortality was already low in the premodern period, and it was as late as in the interwar period that the continuous mortality decline started. Figure 5.5 shows the percentage of "category I" in the basic classification of causes of deaths in vital statistics of Japan. Infectious diseases were combined with general illnesses such as cancer, rheumatism, diabetes, and beriberi until 1922. Since 1923, these two groups were separated and the category I was renamed "infectious and epidemic diseases," causing a serious discontinuity. Although the category I was renamed again as "infectious and parasitic diseases," there was no essential change in the grouping of causes. The changes were to adjust to revisions of the International Classification of Disease (ICD) defined by the International Statistical Institute. In any case, the share of infectious disease did not decline in modern Japan despite the decrease in deaths from smallpox and cholera.

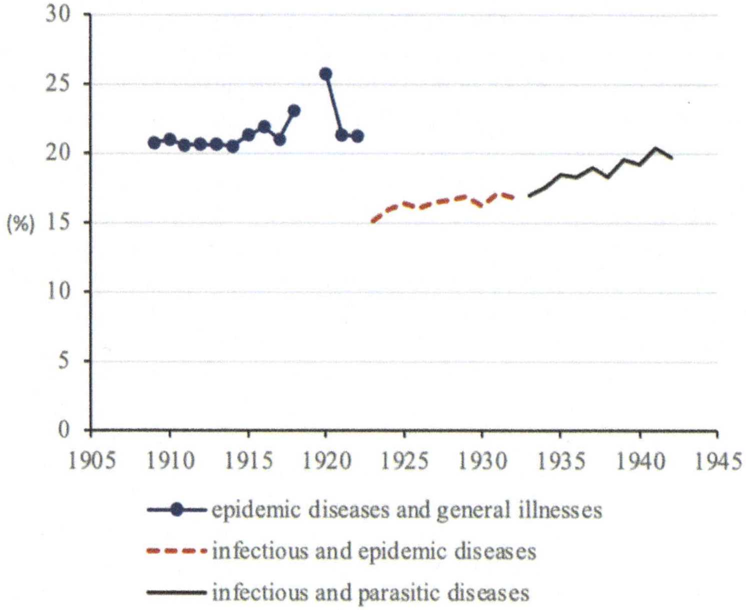

Fig. 5.5 Share of infectious diseases in total deaths: Japan 1909–42

Figure 5.6 shows the shares cholera, dysentery, typhoid, smallpox, tuberculosis, and influenza in the total deaths in Japan. It is apparent that the increase in deaths from infectious diseases was caused by the tuberculosis. Although the tuberculosis existed in Japan since the ancient period, its prevalence drastically increased during the Meiji period. Young women in the textile industry and young men in the military service were most susceptible to the disease. Those who developed the symptom and could not continue to work were sent back to the homeland, encouraging the prevalence in rural areas. The tuberculosis accounted for 60% of deaths of people aged 15–34 years, and called the "national disease" (Ishi 2018). It is understood that the declining competing risk of acute infectious diseases resulted in the increase in tuberculosis that was difficult to prevent and treat. It was after WWII that degenerative diseases replaced both acute and chronical infectious disease and became the major causes of deaths.

The method of variolation was transmitted to Japan in the 1740s and had some effect to prevent the smallpox. Although there were some attempts in the early nineteenth century, it was in the 1840s that the effectiveness of vaccination using the cowpox strain was recognized and widely used in Japan (Fujikawa 1912; Shin 2006). Because the Meiji government implemented the public vaccination program, the death toll from smallpox has never exceeded 10,000 since 1898. On the other hand, the threat of measles remained, and more than 20,000 persons died in 1943. Although there were epidemics of dysentery in the 1880s and 1890s, the disease decreased owing to such public health formulae as the Law of Port Quarantine, the

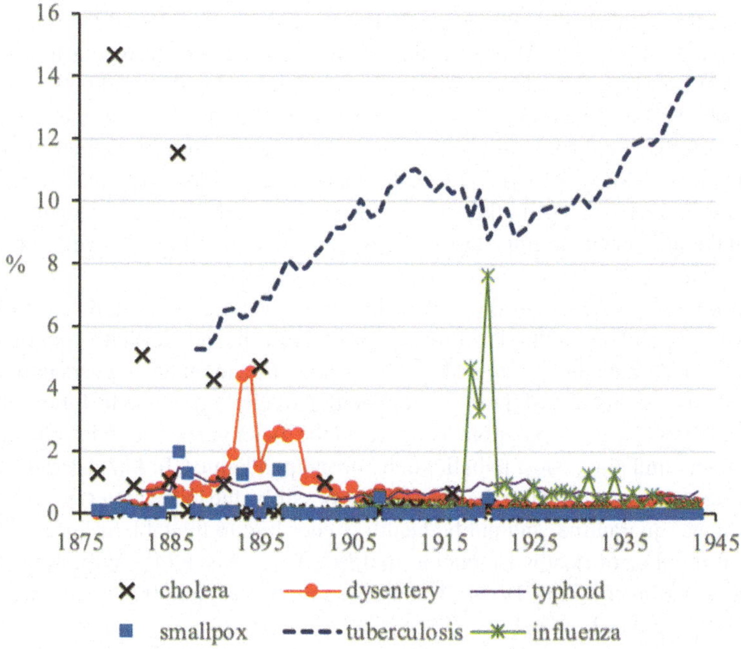

Fig. 5.6 Share of specific diseases in total deaths: Japan 1877–1943. Ito (1995), Vital Statistics

Law of Waste Disposal, and the Sewerage Law enacted in 1899–1900 (Ito 1995). The typhoid was more difficult to prevent and killed more people than the dysentery until the 1940s.

As already mentioned, cholera prevailed in 1821–22, 1858, and 1862 during the Edo period. After the Meiji Restoration, the death toll of cholera exceeded 1000 in 1877, 1879, 1881–82, 1885–86, 1890–91, 1895, 1902, 1907, 1910, 1912, 1916, and 1920. As Fig. 5.6 suggests, the damages in 1879 and 1886 were serious, killing more than 100,000 people. The pestilence in 1877–79 occurred during the fourth worldwide pandemic of cholera in 1863–79. The prevalence started in 1877 from Yokohama and Nagasaki, possibly from different trade ships. This pandemic had a great impact on the quarantine policy of the Meiji government, and the Pestilence Preventive Rule was enacted in 1880. However, the strict management by police caused a harsh antipathy of people and resulted in riots in various areas (Sugiyama 1995). Cholera was transmitted to Korea in 1877 and proceeded to the north from Busan (Shin 2006). During the fifth worldwide pandemic in 1881–96, cholera hit Japan in 1881–82, 1885–86, 1890–91, and 1895. In 1894, cholera was transmitted to Korea by the Japanese soldiers returning from Manchuria, and more than 10,000 Koreans died (KSID 2009).

Although cholera became less lethal in Japan in the twentieth century, its damage was still large in Korea and Taiwan. Table 5.20 summarizes the death toll of cholera in Japan, Korea, and Taiwan in the early twentieth century taken from Iijima (2009).

In Japan, the number of deaths from cholera never exceeded 1% of the total deaths in the twentieth century. However, the deaths from cholera in Korea in 1919–20 accounted for about 2% of the total deaths assumed to be about 600,000. Because of the March 1st Independence Movement that broke out in 1919, the Governor-General's Office could not effectively cope with the concurrent epidemics of cholera and smallpox (KSID 2009). In Taiwan, the deaths from cholera explained about 2.7% of all deaths in 1919. While the cholera raised the number of deaths by more than 10% in the nineteenth-century Japan, the damage was possibly even larger in Korea and Taiwan.

The plague pandemic in the end of the nineteenth century started from Hong Kong in 1894 and transmitted to Taiwan, Japan, Hawaii, and Northern America. The plague landed Taiwan in 1896, and 23,508 people died in eighteen years until 1913. Although the death toll widely overwhelmed that of cholera shown in Table 5.20, the plague did not cause a population decline as in the medieval period. It is apparent that both cholera and plague lost lethality in recurrent pandemics. In Japan, the death toll from plague was even smaller than in Taiwan, possibly due to the better infrastructure of medicine, quarantine, and public health. According to the table in Iijima (2009), the total number of deaths by plague in 1897–1913 was 1,245. While the plague prevailed in Manchuria in 1910–11 and killed more than 50,000 people, there was no case reported in Korea.

Although the deaths from influenza presented in the vital statistics of Japan in 1918–20 sum up to 220,000, the Ministry of Interior estimated that 389,000 people died by the Spanish flu pandemic (Hayami and Koyama 2004). The latter accounted for 9.3% of the total deaths in 1918–20, which was still smaller than the damage of cholera in 1879 and 1886. According to the Ministry of Interior, the death tolls in Korea and Taiwan were 185,000 and 44,000, respectively. These explained 10.3% of the total deaths in Korea and 12.8% in Taiwan. Thus, unlike against the cholera and the plague, Japan was as vulnerable against the Spanish flu as its colonies.

Although malaria prevailed in whole Eastern Asia, it had an especially strong impact in Taiwan and produced notoriety of an unhealthy island. The Dutch Council of Formosa reported a population decline due to epidemics including malaria in1653. When Koxinga landed Taiwan in 1661, 70–80% of soldiers fell ill and many of them died. After the Qing dynasty annexed Taiwan, Chinese officials were aware of the island's unhealthy climate. When Japan and the Qing confronted in Taiwan in 1874, 550 Japanese soldiers and 2,000 Chinese soldiers died of illness. During the attempt of occupying Taiwan in 1895, more Japanese soldiers died of illness than the violence in the battle. The prevalence of malaria continued after the annexation and was the primary cause of deaths in Taiwan until 1916 (Liu and Liu 2001). The Governor-General's Office strictly promoted the measures to cope with the malaria parasite Plasmodium through the paojia system. Inhabitants in malaria-endemic areas were forced undergo blood tests, and those who were affected had to be dosed with quinine. The Malaria Quarantine Rule in 1913 defined that non-cooperation without a reason was punished with a fine (Wakimura 2002; Iijima 2009). Figure 5.7 shows the shares of important diseases in Taiwan calculated from the vital statistics. The public health measures promoted by the Japanese was effective, and the deaths from

Table 5.20 Death toll from Cholera: 1900–30

	Japan	Korea	Taiwan
1902	8,012		613
1903	139		0
1904	48		1
1905	34		0
1906	29		0
1907	1,702		2
1908	297	101	0
1909	158	1,466	0
1910	1,656	382	8
1911	35	2	0
1912	1,763	78	256
1913	106	1	0
1914	100	0	0
1915	63	1	0
1916	7,482	1,254	16
1917	718	1	1
1918	32	0	1
1919	356	11,533	2,693
1920	3,417	13,568	1,675
1921	35	1	0
1922	542	23	0
1923	31	0	0
1924	?	0	0
1925	363	5	3
1926	13	159	11
1927	3	0	0
1928	1	0	0
1929	114	15	0
1930	1	0	0

Iijima (2009)

malaria decreased significantly. The pneumonia and bronchitis replaced malaria as the primary cause of deaths in 1917, and malaria's rank declined to the tenth in 1935. The eradication of malaria apparently contributed to the mortality decline in Taiwan (Chen 1979).

Unlike in Japan, the share of all infectious diseases in the total number of deaths decreased because the tuberculosis did not increase notably. Shepherd (2011) pointed out that the decline in tuberculosis death rates in 1939–41 reflected underreporting caused by the regulation enacted in 1938. Many Taiwanese asked of doctors to report

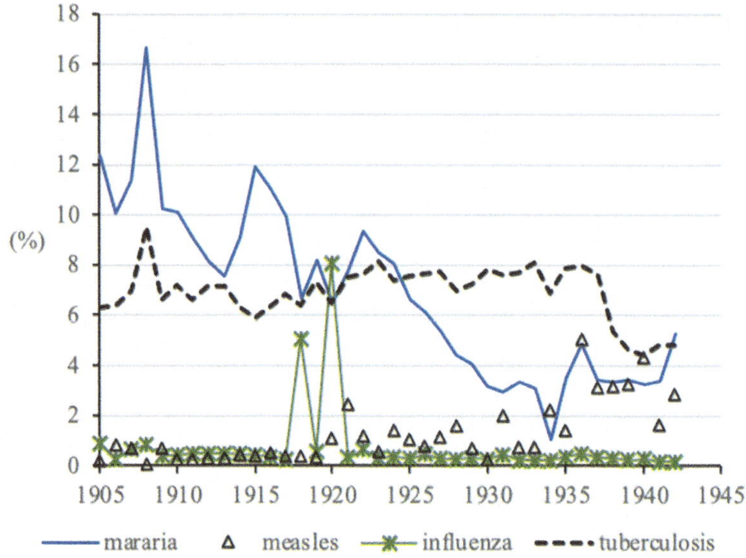

Fig. 5.7 Share of specific diseases in total deaths: Taiwan 1905–42. Vital Statistics

deaths as due to causes other than tuberculosis to avoid quarantine and cremation. It is more likely that the share of tuberculosis remained around 8% after 1939. The reasons for the low tuberculosis prevalence in Taiwan may include the warm weather, the open residential environment, the slow urbanization, and low daily mobility that prevented the invasion of tuberculosis into rural areas. On the other hand, Koreans suffered more from the disease than the Japanese. The Police Department of Korea estimated in 1941 that there were approximately 40,000 annual deaths from the tuberculosis, although only 4,000 were registered (KSID 2009). According to Shin (2006), the morbidity rate of tuberculosis in Korea was 28% in 1934, which was more than twice in Japan and was the third highest in the world. If this was true, it is likely that the share of infectious and parasitic diseases increased during the Japanese era, as in Japan.

The deaths from smallpox and dysentery decreased significantly in Taiwan but those from measles and typhoid did not. The smallpox was prevalent at the time of annexation but decreased dramatically after the vaccination was enforced in 1903 (Chen 1979). However, as suggested in Fig. 5.7, the deaths from measles increased after the mid 1930s. In 1936, the epidemic of measles killed 5,280 persons, accounting for 5% of the total deaths. While more than 100 died for typhoid every year, the annual deaths from dysentery never exceeded 100 after 1916. The smallpox vaccination began at the end of the nineteenth century in Korea, and Japan forced it since 1905 with a threat by weapon. Although its impact was not impressive during the early period, the deaths from smallpox notably decreased in the 1930s (Shin 2006). It is difficult to know the trend of each disease because of the incomplete death registration in colonial Korea. However, KSID (2009) evaluated that infectious diseases other than the smallpox and cholera did not decrease significantly. The measles was already a childhood disease and many children died with a complication of pneumonia or bronchitis. The typhoid and dysentery prevailed every summer and killed hundreds of Koreans.

In the nineteenth-century China, cholera prevailed in 1820, 1840, 1858, 1862, and 1877 (Iijima 2009). There is no record during the fifth worldwide pandemic in 1881–96. In Japan, the deaths from cholera exceeded 100,000 in 1879. If the excess death rate was equivalent in the 1877 epidemic in China, there were almost one million deaths from cholera. In May 1919, the cholera landed Guangdong province and proceeded northward to Fujian, Zhejiang, Shandong, and Liaoning provinces. In August, cholera penetrated inland areas via the railway to Harbin and Vladivostok. In Korea, there were about 25,000 deaths from cholera in 1919–20. If the damage was same in China, applying the ratio of Korean population to the Chinese, the death toll in China was 25,000 * 25 = 625,000. According to Iijima, there were 50,000 deaths from plague in Manchuria in 1910–11. Although the death toll of plague in China before 1910 is unknown, Iijima asserted that the demographic impact was considerably smaller than cholera. The deaths from Spanish flu in China has been estimated to be between 4.0 and 9.5 million (Wakimura 2002). However, if we apply the ratio of populations in Japan and China to the death toll in Japan, we obtain a smaller estimate of 3.2 million. Because Japan was as vulnerable as Korea or Taiwan, there is no reason to assume that the damage was significantly larger in China.

While there was no great famine within the Japanese empire in the modern period, the 1942 Henan famine in China is notorious as a man-made disaster (Woo 2016). The Japanese army occupied important cities such as Shanghai, Nanjing, and Suzhou in 1938, and advanced to Zhengzhou in Henan province. In June, the nationalist leader Chiang Kai-shek ordered an intentional destruction of dikes on the Yellow River to stop the invasion by Japan. While the Japanese army suffered little damage, 12.5 million neighboring inhabitants were victimized and 890,000 were dead by the man-made flood. The agriculture of Henan suffered a heavy damage, and newly made swamps and marshes worked as a hotbed of locus plagues. In the autumn of 1942, the harvest was ruined by a massive locust outbreak, and the province was covered with starvation, cannibalism, and human trafficking of children. However, the local government neither reported the situation to Chongqing nor attempted a relief activity. It was only after the Westerners' concern was raised by a report by an American journalist in February 1943 that the nationalist government began to support victims in Henan. According to Woo, there were five million excess deaths throughout the province.

It is thought that WWI was the first "total war" in which the full national power was spent accompanying huge victims. Before WWI, a war was relatively minor factor of mortality unless it accompanied a famine or an epidemic. The death toll of 14,000 in the First Sino-Japan War was so small that Takase (1991) did not consider in his backward population projection. Although he considered 85,000 deaths in the Russo-Japanese War, it was smaller than the deaths in the Spanish flu (389,000) and the Great Kanto earthquake (105,000). The Japanese military forces killed more than 10,000 enemies in the First Sino-Japan War including the invasion of Taiwan, the

Boxer Rebellion, and the Russo-Japanese War. These did not accompany a famine or an epidemic, and damages were smaller than the Taiping Rebellion. While WWI was a great catastrophe in Europe killing ten million people, battles in Eastern Asia were of small scale. Although there were a huge number of deaths in the second Sino-Japan War (1937–45), the estimations of death tolls range in a wide range. According to the Wikipedia (seen on 30 July 2023), the deaths of the Chinese ranged between 266,800 and 1,000,000. As for the Japan side, it is estimated that more than 2,121,000 soldiers and 1,000,000 civilians died in battles against China and the US.

The age structure in 1950 was calculated from UNPD (2022) and shown in Fig. 5.8 to examine the impact of WWII. The age structure of Korea was that of the combined population of the south and the north. Germany was added for comparison. In Germany, the cohort born in 1915–20 and aged 25–29 years on 1 July 1945 suffered most from the war. No Eastern Asian country had such a singular age pattern in 1950. The proportion of the 1915–20 cohort ranged in a narrow interval between 6.1% in Korea and 7.2% in Taiwan. Figure 5.9 shows the sex ratio by age group in 1950 also based on UNPD (2022). In Germany, the sex ratio was below 80% in cohorts born in 1910–30 and aged 20–39 at the midyear of 1945, implying there were excess deaths of males due to the war. The same pattern can be seen in Japan, although the magnitude was smaller than in Germany. The low sex ratio of young population in Korea is incomprehensible. Considering that the Korean War broke out on 25 June 1950, the midyear of 1950 is too early to show the impact of the war. It is suspected that the UN demographers conducted a backward projection from the 1955 census without considering the date of outbreak. Taiwan's high sex ratio in young and middle-aged population is acceptable because it was in 1949 that one million nationalists retreated to Taiwan. As far as we rely on the estimation by UNPD, no impact of war was found for the population in China.

5.9 Modern Household Changes

In Table 5.21, the mean household sizes in regions of the Japanese Empire are author's calculation by dividing the total population with the number of all households including institution in censuses. Figures for the Republic of China are from the population census in 1912 and the household registration survey in 1936–37 cited by Minami and Makino (2014). As mentioned above, the mean household size in Tokugawa Japan was usually less than five persons. Although there was no big change until the 1920s, the mean size increased in 1925–35. Although the mean household size in Korea was larger, the trend was similar to that of Japan. The Taiwanese households were further larger, and the mean size continued to grow after 1935. There was little change in the mean household size in the Republic of China.

The causes of an increase in the mean household size can be divided into the growth of child population and the decline of headship rate in the adult population. Because a child under the age 15 rarely takes the headship, an increase in the proportion of child population directly results in the increase in the mean household size. On the

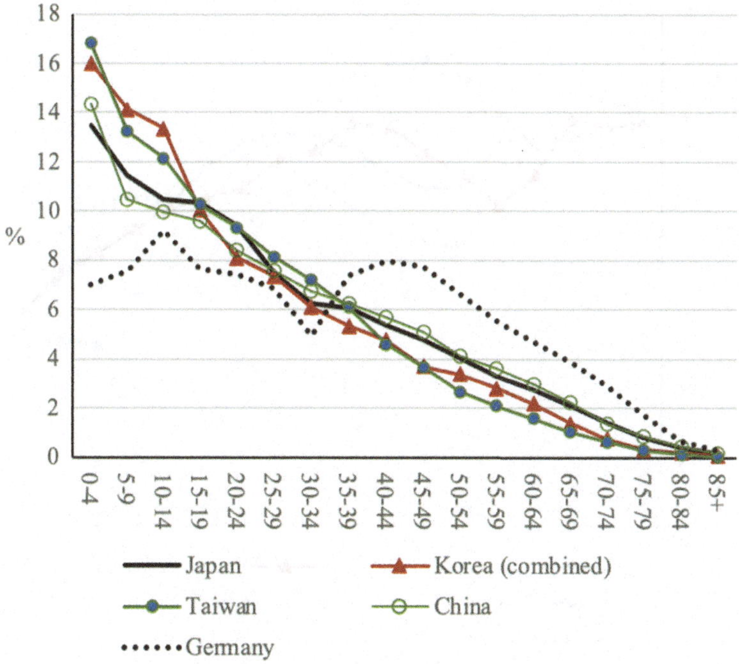

Fig. 5.8 Age structure in 1950. Based on UNPD (2022)

other hand, if the headship rate declined due to the delayed marriage or the improved survivorship of parents, the mean household size would be larger. By contrast, a rise in the headship rate owing to the value change encouraging the intergenerational independence would make the household size smaller. The mean household size h can be expressed as $h = h\,(c + a)$, where c and a are the proportion of child population and that of adult population, respectively. Then, the change in h can be decomposed as $h_2 - h_1 = (h_2\,c_2 - h_1\,c_1) + (h_2\,a_2 - h_1\,a_1)$.

Table 5.22 shows the decomposition results. In Japan, while both the child population and the adult population promoted the expansion of households in 1925–35, the decline in the adult headship rate had slightly lager impact. In Korea, the rise in the average household size was entirely attributed to the growing child population. The child population had a dominant impact also in Taiwan, although the adult headship rate showed a little contribution. In Japan, the proportion of child population decreased from 36.9% in 1935 to 36.7% in 1940, and the headship rate increased, resulting in the decline in the average household size. In Korea, the proportion of child population continued to grow from 40.9% in 1935 to 41.9% in 1940. However, the average household size declined owing to the increased adult headship rate. Although the decomposition for Taiwan cannot be shown here, it is likely that the expansion of households in 1935–40 was also caused by the growing child population as in 1925–35.

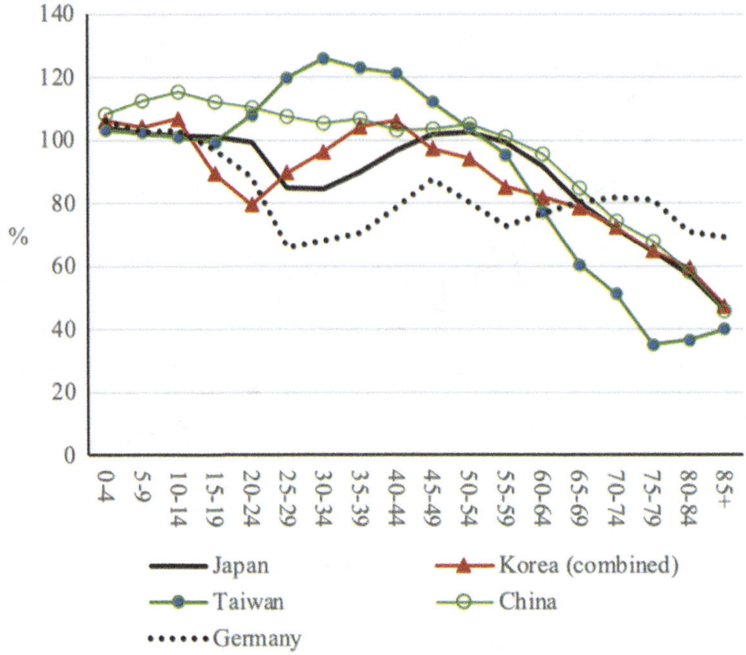

Fig. 5.9 Sex ratio by age in 1950. Based on UNPD (2022)

Table 5.21 Mean household size: 1920–40

	Japan	Korea	Taiwan	China
1920	4.99		5.30	5.37[a]
1925	4.98	5.25	5.50	
1930	5.07	5.28	5.68	
1935	5.13	5.33	5.81	5.38[b]
1940	5.02	5.30	5.85	

Census of Japan, Korea, and Taiwan; Minami and Makino (2014)
[a] 1912
[b] 1936–37

5.10 Summary and Conclusion

Japan was successful in linking protoindustrialization in the early nineteenth century to industrialization by introducing the English model of production. Thus, the economic growth of Japan could support population growth at more than 1% of annual rates. The population growth of Korea and Taiwan probably started in the late nineteenth century immediately before the annexation to Japan and accelerated during the Japanese rule. On the other hand, the population growth of China began

Table 5.22 Decomposition of mean household size: 1925–40

	Japan	Korea	Taiwan
(1925–35)			
Chenge in mean household size	0.1500	0.0804	0.3107
Due to child population	0.0645	0.0974	0.2553
Due to adult population	0.0854	– 0.0170	0.0554
(1935–40)			
Chenge in mean household size	– 0.1125	– 0.0241	
Due to child population	– 0.0520	0.0393	
Due to adult population	– 0.0605	– 0.0634	

too early to learn how to break through the limits of agrarian society. China seems to be the first country on the Eurasian continent that achieved a new epidemiological balance and realized population and economic growth in the eighteenth century. Since it was earlier than in England, growth was checked by the limitations on land, and the population collapsed in the 1850s and 1860s. Although the situation improved after the 1870s, population and economic growth rates were significantly lower than in the Japanese Empire until the foundation of the People's Republic of China.

Zelinsky's thesis that the impact of urbanization appears after the peak of frontier migration was supported by the distributional changes in Japan, Korea, and Taiwan. While the concentration on urban areas became apparent only after 1920 in Japan, the equalization due to frontier migration continued throughout the modern period in Korea and Taiwan. The difference in agricultural competitiveness between Korea and Taiwan brought about different destinies of population. Because the rice monoculture economy in Korea could not absorb the surplus population, massive out-migration toward cities and abroad continued, and Korean villages remained underdeveloped. To the contrary, various cash-crops were planted in Taiwanese villages and accumulated wealth for the rural industrialization. Taiwanese villagers had little motivation to out-migrate and showed notably lower propensity to move than the Japanese or Koreans.

There were huge outflows of the Japanese and Korean populations because of the expansion of the Japanese empire and the high population pressures in homelands. The population of overseas Japanese inhabiting in overseas colonies and battlefields exceeded six million at the end of WWII. The population of overseas Koreans mainly inhabiting in Japan and Manchukuo also exceeded two million. To the contrary, Taiwan was an almost closed population during Japan's rule. Although the population of overseas Chinese was as large as 7.5 million in 1935, the ratio to the homeland population was only 1.45%.

The rapid population growth during the modern period was caused by rising fertility in Japan and by declining mortality in Korea and Taiwan. While Korean women and Chinese women married very early, the female age at marriage in 1930 was relatively high in Japan and Taiwan. Then, fertility was possibly higher in Korea and China than in Taiwan. Mortality of Japan was already low in the premodern period, and further declined during the interwar period. Mortality decline was faster in Taiwan where the agricultural economy was better than in Korea. Although mortality declined in the Japanese empire, the epidemiological transition of main causes of deaths from infectious to degenerative diseases did not begin Japan and Korea until the end of WWII. The deaths from tuberculosis drastically increased and the disease was called the "national disease." In Taiwan, however, the increase of tuberculosis was moderate, and the share of all infectious and parasitic diseases possibly declined. As far as we rely on the estimation by the UNPD, the demographic impact of WWII was small in Eastern Asia and mostly absent in China.

The household size became larger because of the increasing number of children living together. Thus, such household changes as the decline in the mean household size and the rise in the propensity to live alone were left for the contemporary period in addition to the fertility transition, the epidemiological transition, the acute urbanization, the delay in marriage, and the increase in eventually unmarried and childless people.

References

Barclay, Gerorge W. 1954. *Colonial Development and Population in Taiwan*. Princeton University Press.

Barclay, George W., Ansley J. Coale, Michael A. Stoto, and T. James Trussell. 1976. A reassessment of the demography of traditional rural China. *Population Index* 42 (4): 606–635.

Cao, Shuji. 2001. *Zhongguo Renkoushi 5 Qing Shiqi*. Fudan University Press (in Chinese).

Cha, Myung-Soo. 2009. Josen hugi-eui chulsanryeok, samangryeok mich ingu jeungga: ne jokpo-e natanan 1700–1899nyeongan saengmok girok-eur iyonghan yeongu. *Korea Journal of Population Studies* 32 (1): 113–137 (in Korean).

Cha, Myung-Soo. 2014. *Gia-wa Gijeok-eui Giweon – Hanguk Gyeongjesa, 1700–2010*. Haeam (in Korean).

Chang, Se-Hun. 2002. Doshihoa. In *Hanguk-eui Ingu*, eds. Kim Doo-Sub, Park Sang-Tae and Eun Ki-Soo, 495–523. National Statistics Office (in Korean).

Chen, Shao-Sheng. 1979. *Taiwan-de Renkou Bianqian yu Shehui Bianqian*. Lianjing (in Chinese).

Chen, Laixing. 2013. Imin-to kakyo. In *Chugoku Keizaishi*, ed. Okamoto Takashi, 232–234. Nagoya University Press (in Japanese).

Cumings, Bruce. 1997. *Japanese Colonialism In Korea: A Comparative Perspective*. Walter H. Shorenstein Asia-Pacific Research Center, Stanford University. https://aparc.fsi.stanford.edu/publications/japanese_colonialism_in_korea_a_comparative_perspective (seen on 8th August 2018)

Cumings, Bruce. 2005. *Korea's Place in the Sun: A Modern History, Updated Edition*. W. W. Norton & Co.

Diamond, Jared. 1997. *Guns, Germs, and Steel: The Fates of Human Societies*. W. W. Norton.

Eckert, Carter J. 1991. *Offspring of Empire, The Koch'ang Kims and the Colonial Origins of Korean Capitalism 1876–1945*. University of Washington Press.

Fujikawa, Yu. 1912. *Nihon Shippei-shi*. Tofudo (in Japanese).

Hajnal, John. 1953. Age at marriage and proportion marrying. *Population Studies* 7 (2): 111–136.

Hayami, Akira, and Miyoko Koyama. 2004. *Taisho Demography – Rekishi Jinkogaku-de Mita Hazama-no Jidai*. Bunshun Shinsho (in Japanese).

Hayami, Akira. 2015. *Japan's Industrious Revolution: Economic and Social Transformation in the Early Modern Period.* Springer.

Ho, Ping-Ti. 1959. *Studies on the Population of China, 1368–1953.* Harvard University Press.

Iijima, Wataru. 2009. *Kansensho-no Chugoku-shi – Koshu Eisei-to Asia.* Chuokoronsha (in Japanese).

Ishi, Yoshikuni. 1972. *Kankoku-no Jinko Zoka-no Bunseki.* Keiso Shobo (in Japanese).

Ishi, Hiroyuki. 2018. *Kansensho-no Sekaishi.* Kadokawa (in Japanese).

Ishida, Shichiro. 1932. *Manshi Keizai Kankei-no Ichikosatsu.* Manshu Bunka Kyokai (in Japanese).

Ishii, Kanji. 1991. *Nihon Keizai-shi.* University of Tokyo Press (in Japanese).

Ito, Shigeru. 1995. Kindai nihon-no toshi-ni okeru sippei-to jinko. In *Koza Bunmei-to Kankyo 7. Jinko, Sippei, Saigai*, eds. Hayami Akira and Machida Hiroshi, 183–196. Asakura Shoten (in Japanese).

Kajitani, Kai, and Kashima. 2013. Kingendai – kokumin kokka keisei-no ayumi-to keizai hatten. In *Chugoku Keizaishi*, ed. Okamoto Takashi, 237–297. Nagoya University Press (in Japanese).

Kasuya, Kenichi, Masato Namiki, and Yusuke Hayashi. 2016. *Gendai Chosen-shi.* Yamakawa Shuppansha (in Japanese).

Kikakuin (The Planning Board). 1939. *Kakyo-no Kenkyu.* Matsuyamabo (in Japanese).

Kim, Cheol. 1965. *Kankoku-no Jinko–to Keizai.* Iwanami Shoten (in Japanese).

Kim, Tu-hŏn. 2008. *Kankoku Kazoku Seido-no Kenkyu.* Translated by Kim Youngmi, Kim Hyang-Nam, and Kim Jung-Nim, Hosei University Press (in Japanese).

Kim, Doo-Sub, Park Sang-Tae, and Eun Ki-Soo (eds.). 2002. *Hanguk-eui Ingu.* National Statistics Office (in Korean).

Kita, Kazuo. 1944. *Manshu Kaitakuron.* Meibundo (in Japanese).

Kito, Hiroshi. 2007. *Zusetsu: Jinko-de Miru Nihonshi – Jomon Jidai–kara Kinmirai Shakai-made.* PHP Interface (in Japanese).

Kobayashi, Fumio. 1942. *Kinsei Shina Keizaishi Kenkyu.* Kobundo (in Japanese).

Komine, Kazuo. 2014. *Manshu – Manchuria-no Kigen, Shokumin, Haken.* Kodansha (in Japanese).

Kondo, Yasuo. 1942. *Manshu Nogyo Keizairon.* Nihon Hyoronsha (in Japanese).

Korean Society of Infectious Diseases (KSID). 2009. *Hanguk Gamyeombyeongsa.* Koonja Publishing Inc. (in Korean).

Kwon, Tai-Hwan, and Kim Doo-Sub. 2002. *Ingu-eui Ihae.* Seoul National University Press (in Korean).

Kwon, Tai-Hwan, Hae-Young Lee, Yunshik Chang, and Eui-Young Yu. 1975. *The Population of Korea.* Population and Development Studies Center, Seoul National University.

Kwon, Tai-Hwan. 2002. Gukjeidong-gwa haiwoe haninsahoe (in Korean), In *Hanguk-eui Ingu*, eds. Kim Doo-Sub, Park Sang-Tae, and Eun Ki-Soo, 153–188. National Statistics Office.

Lee, Yong-Hoon. 2009. *Daikan Minkoku-no Monogatari.* Translated by Nagashima Hiroki, Bungei Shunju (in Japanese).

Liu, Ts'ui-Jung, and Shi-Yung Liu. 2001. Disease and mortality in the history of Taiwan. In *Asian Population History*, eds. Liu Ts'ui-jung, James Lee, David Sven Reher, Saito Osamu, and Wang Feng, 248–269. Oxford University Press.

Maddison, A. 2001. *The World Economy: A Millennial Perspective.* Development Centre of the OECD.

McNeill, William H. 1976. *Plagues and Peoples.* Anchor.

Minami, Ryoshin, and Makino, Fumio. 2014. *Asia Choki Keizai Tokei 3 Chugoku.* Toyo Keizai Shimposha (in Japanese).

Mizoguchi, Toshiyuki. 2008. *Asia Choki Keizai Tokei 1 Taiwan.* Toyo Keizai Shimposha (in Japanese).

Mizuno, Naoki, and Kyung-Soo Moon. 2015. *Zainichi Chosenjin – Rekishi-to Genzai.* Iwanami Bunko (in Japanese).

Mizushima, Tsukasa, Kubo Toru, and Shimada Ryuto. 2015. *Asia Keizaishi Kenkyu Nyumon.* Nagoya University Press (in Japanese).

Nakamura, Satoru. 2004. Shono keiei-no kikakushiteki kento. In *Nihon Shihonshugi-to Chosen,Taiwan – Teikokushugika-no Keizai Hendo*, eds. Hori Kazuo and Nakamura Satoru, 77–93. Kyoto University Press (in Japanese).
National Institute of Population and Social Security Research (IPSS). 2023. *Jinko Tokei Shiryoshu 2023 Kaiteiban* (in Japanese).
Obuchi, Hiroshi, and Morioka, Hitoshi. 1981. *Keizai Jingogaku*. Sinhyoron (in Japanese).
Statistics Bureau, Cabinet Office. 1930. *Meiji Gonen Iko Wagakuni-no Jinko*, Chosa Shiryo Dai 3-shu (in Japanese).
Okamoto, Takashi. 2013. Minshin – dento keizai-no keisei-to hensen (15–19 Seiki). In *Okamoto Takashi*, ed. Chugoku Keizaishi, 177–236. Nagoya University Press (in Japanese).
Okazaki, Yoichi. 1986. Meiji Taishoki-ni okeru nihon jinko-to sono dotai. *Journal of Population Problems* 178: 1–17 (in Japanese).
Omran, Abdel R. 1971. The epidemiological transition. *Milbank Memorial Fund Quarterly* 49 (4): 509–508.
Oshima, Hisayuki. 2015. Chugokujin rodosha-no donyu-to rodo shijo. In *Shokuminchi Taiwan-no Keizai Kiban-to Sangyo*, ed. Sunaga Noritake, 249–267. Nihon Keizai Hyoronsha (in Japanese).
Park, Keong-Suk. 2009. Shigminchi shigi (1910nyeon~1945nyeon) Joseon-eui ingu dongtae-wa gujo. *Korea Journal of Population Studies* 32 (2): 29–58 (in Korean).
Park, Jae-Il. 1957. *Zainichi Chosenjin-ni kansuru Chosa Kenkyu*. Shinkigensha (in Japanese).
Park, Yi-Taek. 2008. Shikminji Shigi Joseonin Ingu-eui Jaeheomtu: 1910–1940. *Daedong Munhwa Yeongu* 63: 331–373 (in Korean).
Park, Kyoung-Ock. 2015. *Kindai Chugoku Tohoku Chiiki-no Chosenjin Imin-to Nogyo*. Ochanomizu Shobo (in Japanese).
Rinji Taiwan Kyukan Chosakai. 1903. *Rinji Taiwan Kyukan Chosakai Daiichibu Chosa Hokokusho, Daiikkai Hokokusho* (in Japanese).
Saito, Osamu. 1997. Asia jinko-shi tenbo. *Keizai Kenkyu* 48 (1): 59–79 (in Japanese).
Saito, Osamu. 2001. Kindai jinko seicho. In *Rekishi Jinkogaku-no Frontier*, eds. Hayami Akira, Kito Hiroshi, and Tomobe Ken'ichi, 67–89. Toyo Keizai Shimposha (in Japanese).
Shepherd, John R. 2011. Trends in mortality and causes of death in Japanese colonial period Taiwan. In *Death at the Opposite Ends of the Eurasian Continent: Mortality Trends in Taiwan and the Netherlands, 1850–1945*, eds. Theo Engelen, John R. Shepherd, and Yang Wen-shan, 45–79. Aksant Amsterdam.
Shin, Kyu Hwan. 2006. *Jilbyeong-eui Sahoesa: Dong-Asia Euihak-eui Jaebalgyeon*. Sallim Books (in Korean).
Shiode, Hiroyuki. 2015. *Ekkyosha-no Seiji-shi: Asia Taiheiyo-ni okeru Nihonjin–no Imin–to Shokumin*. Nagoya University Press (in Japanese).
Sugihara, Kaoru. 1994. Kakyo-no imin network-to tonan Asia keizai – 19-seiki-matsu ~ 1930-nendai-wo chushin-ni. In *Choki Shakai Hendo: Asia-kara Kangaeru*, eds. Mizoguchi Yuzo, Hamashita Takeshi, Hiraishi Naoaki, and Miyajima Hiroshi, 163–195. Tokyo University Press (in Japanese).
Sugiyama, Shinya. 1995. Bakumatsu kaiko-to ekibyo – ibunka sesshoku-to jinko. In *Koza Bunmei-to Kankyo 7. Jinko, Sippei, Saigai*, eds. Hayami Akira and Machida Hiroshi, 166–182. Asakura Shoten (in Japanese).
Takaoka, Kumao, and Uehara Tetsusaburo. 1943. *Hokushi Imin-no Kenkyu*. Japan Society for the Promotion of Science (in Japanese).
Takase, Masato. 1991. 1890–1920nen-no wagakuni-no jinko dotai-to jinko seitai. *The Journal of Population Studies* 14: 21–34 (in Japanese).
Ueda, Makoto. 1995. Shinchoki chugoku-no saigai-to jinko. In *Koza Bunmei-to Kankyo 7. Jinko, Sippei, Saigai*, eds. Hayami Akira, and Machida Hiroshi, 236–248. Asakura Shoten (in Japanese).
United Nations Population Division. 2022. *World Population Prospects, The 2022 Revision*.
Wakatsuki, Yasuo. 1995. *Shinpan Sengo Hikiage-no Kiroku*. Jijihyoronsha (in Japanese).

Wakimura, Kohei. 2002. *Kiga, Shippei, Shokuminchi Tochi – Kaihatsu-no Nakano Eiryo Indo.* University of Nagoya Press (in Japanese).
Woo, Seungyeop. 2016. *Daegigeun-i Onda – Sesang-eul Bakkun Gamum-gwa Gigeun-eui Yeoksa.* Cheom Books (in Korean).
Yamanaka, Mineo. 2005. Manshukoku jinkoutoukei-no suikei. *Journal of Tokyo Keizai University* 245: 167–190 (in Japanese).
Yasukawa, Masaaki. 1977. *Jinko-no Keizaigaku.* Shunjusha (in Japanese).
Zelinsky, Wilbur. 1971. The hypothesis of the mobility transition. *Geographical Review* 61 (2): 219–249.
Zensho, Eisuke. 1927. *Chosen-no Jinko Gensho.* Governor-General of Korea (in Japanese).
Zhao, Wentan, and Xie, Shujun. 1988. *Zhongguo Renkoushi.* Renmin Chubanshe. (in Chinese)

Chapter 6
Population in Contemporary Eastern Asia

6.1 Upheavals and Developments

Japan was defeated in the Pacific War and was occupied by the Allied Powers until 1952. With the demolished infrastructure and a huge wave of repatriation, food supplies and unemployment were terribly serious problems. However, the Japanese economy was boosted by wartime demands during the Korean War (1950–53). At the same time, a concern with overpopulation prompted a drastic fertility decline in the 1950s. After the Korean War, economic growth was further accelerated by growing investments, a reinstated infrastructure, and improved purchasing power. The miraculous economic growth of Japan was disrupted by the transition to the floating exchange rate system and the oil crisis in the early 1970s. Although the growth rate slowed, the Japanese economy sustained stable growth owing to the local development program under Tanaka Kakuei's cabinet (1972–74), growing exports of automobiles and home electronics since the late 1970s, and the bubble boom of the late 1980s. However, the bubble economy collapsed in the early 1990s, and Japan has since suffered from slow economic growth. Demographically, Japan's TFR fluctuated around the replacement level between the mid-1950s and the mid-1970s. Then, the TFR has never regained the replacement level but recorded the lowest value of 1.26 in 2005. Japan achieved the longest life expectancy worldwide in the 1980s. The combination of low fertility and extremely low mortality made Japan's population oldest in the world by 2010.

After the surrender of Japan, the Korean peninsula was occupied separately by the US and the Soviet Union. Attempts to establish a unified government failed; the Republic of Korea was founded in the south and the Democratic People's Republic of Korea was founded in the north in 1948. Damage from the Korean War was in both sides. The economic growth in the south was slower than the north until the early 1960s. However, the development dictatorship of Park Chung-Hee brought about the dramatic economic success called the "miracle on the Han River." By the late 1980s, it became difficult to sustain the dictatorship, and various democratic reforms

T. Suzuki, *Comparative Population History of Eastern Asia*,
https://doi.org/10.1007/978-981-99-9367-3_6

were achieved during the presidency of Roh Tae-Woo. In 1997, the Asian financial crisis started from Thailand hit Korea and the Kim Young-Sam government had to ask the financial aid from the IMF. Although the Korean economy recovered quickly from the damage, growth has been slower than before the crisis, and US$ 30,000 per capita income was not achieved until 2017. Korea recorded the TFR of 0.78 in 2022, possibly the lowest value ever recorded by a country with a large population. It is certain that Korea will replace Japan to become the most aged country in the predictable future.

When Japan surrendered to the Allied Powers in 1945, Taiwan was requisitioned by the Republic of China. The Chinese civil war ended in the victory of the communists, and the KMT retreated to Taiwan in 1949. Although the early KMT administration caused political and economic confusions including the February 28 incident (1947) and the hyperinflation, the Taiwanese economy was less damaged than in Korea where the war broke out. Rapid economic growth began in the early 1950s, and the main drive shifted from agriculture to light industries activated by small and medium-sized enterprises. The KMT government restricted the political participation of the Hoklo and Hakka Chinese, not to mention the aborigines. The political and governmental sectors were monopolized by the Mainlanders, and dialects other than Mandarin were prohibited in official places. Interestingly, democratization was achieved in 1987, the same year as Korea. The KMT could not crush the claim of ethnic majorities and permitted the foundation of the Minjindang (MJD), lifting the martial law. Although Taiwan was not seriously damaged by the Asian financial crisis, the scale merit of Korea, with its large population, heavy industries, and gigantic "chaebols" (family-owned conglomerates) has allowed Korea a more advantageous position in recent years. Korea and Taiwan resembled also in terms of the trajectory of fertility decline and have been suffering lowest-low fertility since the early 2000s.

China was slower to begin economic development than other parts of Eastern Asia. China fought against the UN coalition forces in the Korean War and had a considerable damage. Although the first five-year plan (1953–57) was achieved successfully and the economy began to recover gradually, Mao Zedong's maladministration including the Great Leap Forward (1955–62) and the Cultural Revolution (1966–76) destroyed the condition of economic success. It was in the 1980s that Mao's failure was amended by the economic reforms led by Deng Xiaoping. The economic growth accelerated in the 2000s and the GDP of China overtook that of Japan in 2010 to become the second largest economy worldwide. Although the statistics of China, especially those of fertility, are less reliable than in other parts of Eastern Asia, it is possible that the TFR has been in the lowest-low level at least in recent years.

6.2 Fertility Transition

Figure 6.1 shows the changes in TFR since the late 1940s. Japan's TFR rapidly declined from 4.32 in 1949 to 2.04 in 1957. This fertility transition occurred without an official governmental anti-natal measure, although the revision of Eugenic Protection Law in 1949 that permitted abortions with economic reasons played an indirect role. Possibilities of famine from food shortage and political disorder from unemployment were avoided through economic growth since the 1950s. Throughout the 1950s and 1960s, the economy of Japan continued to grow rapidly, and the TFR fluctuated around the replacement level with an exceptional decline in 1966. According to the sexagenary cycle, this year was "hinoeuma" with a superstition that a girl born in this year would be evil. In any case, the TFR stayed around the replacement level until the mid-1970s. A quarter century from 1950 was the golden age of Japan with the high economic growth, the altruism toward children, and the clear role division between a male breadwinner and a fulltime housewife.

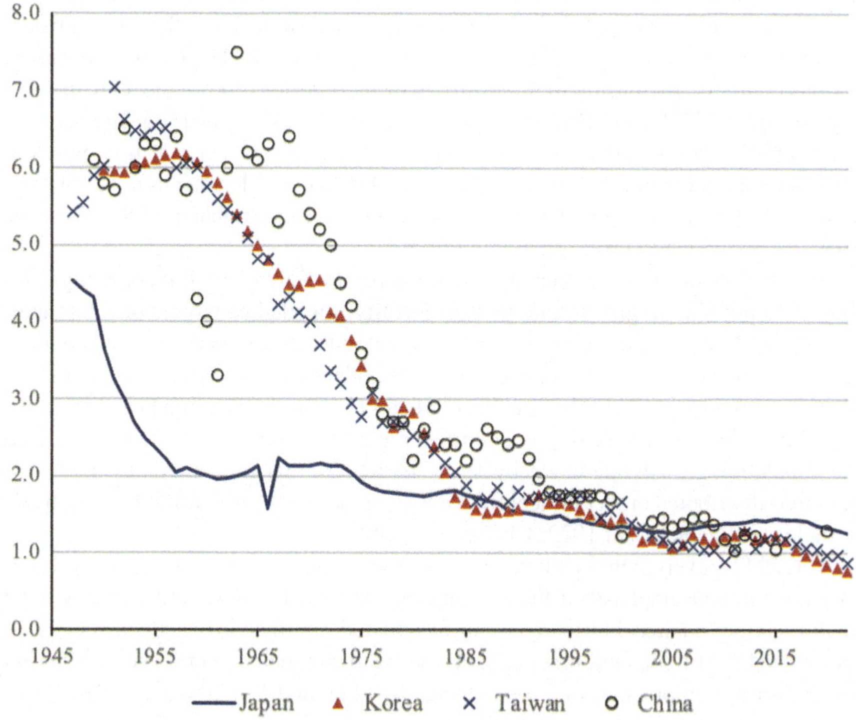

Fig. 6.1 Total fertility rate: 1947–2022 Japan: IPSS (2023a).Korea 1950–69: United Nations Population Division (2022). Korea 1970–2022: Statistics Korea.Taiwan 1947–2008: Ministry of the Interior (2009).Taiwan 2009–22: DGBAS China 1949–89: Wakabayashi (1989).China 1990–99. Tian (2007).China 2000, 2010, 2015, 2020: Census.China 2003–09, 2011–13: China Statistical Yearbook

The second demographic transition (van de Kaa 1987) of Japan started in the late 1970s and the TFR declined to the below-replacement level. As expected in the theory developed by Lesthaeghe and van de Kaa, the process accompanied value changes such as the decreasing importance of religions, the erosion of the familial norm of supporting elderly parents, a reduced preference for sons, and an increasing tolerance toward divorce and premarital intercourse (Ato 1997). The political turning point was in 1989, when Japan's TFR hit its historically low level of 1.57. It was once believed that the TFR of 1.58 in 1966 prompted by hinoeuma superstition was so abnormal that the TFR would never drop below this level. After this "1.57 shock," the government declared to start pro-natal policy to cope with declining fertility. Although various measures including child allowance, childcare leave, childcare services, and wife-work balance campaign were applied, Japan's fertility decline did not stop and the TFR recorded the lowest value of 1.26 in 2005. While Japan escaped lowest-low fertility after 2005, the TFR declined during the COVID-19 pandemic and recorded 1.26 again in 2022.

The governmental family planning program played an important role in fertility decline of Korea, Taiwan, and China. As mentioned in the previous chapter, Japan established a reliable population statistics system in Taiwan. Thus, the vital statistics of Taiwan are reliable except for the late 1940s, when the KMT governance caused serious conflicts. According to the Ministry of the Interior (2009), the TFR in 1951 was as high as 7.04. The TFR value slowly declined to 4.825 in 1965. Japan could not establish a good vital statistics system in Korea, and coverage was not improved after the foundation of the Republic of Korea. The Statistics Korea has not published estimated vital rates before 1970. The estimated and interpolated TFR values for 1950–69 by the UNPD were shown in Fig. 6.1.

The leaders of the developmental dictatorship in Korea and Taiwan recognized that rapid population growth due to high fertility impeded economic development. The Park Chung-Hee government in Korea emphasized the necessity of reducing fertility in the First Five-Year Plan (1962–66) and launched the governmental family planning program in 1961. The Ten-Year Plan for the Family Planning Program established in 1963 aimed at raising the contraceptive practice rate to 45% and reducing the population growth rate to 2% by 1971. To achieve this goal, governmental organizations distributed oral contraceptives and encouraged the use of IUDs, in addition to male sterilization (Ishi 1972; Choi et al. 2005).

The KMT government in Taiwan also started a family planning program in 1964. The government emphasized the advantages of small families, and workers from health stations visited households and distributed contraceptives. Recurrent KAP (Knowledge, Attitude, and Practice) surveys were conducted, and it was shown that the contraceptive practice rate rose constantly (Lee and Lin 2016). As a result of these governmental efforts, in addition to economic development, the replacement level of fertility was achieved by Korea in 1983 and by Taiwan in 1984.

Since Mao Zedong asserted that population growth is good and rejected the necessity of anti-natal policy, the government could not powerfully promote anti-natal policy intervention until his death in 1976. The one-child policy was introduced in 1979, and second births by urban Han Chinese were strictly restricted by a financial penalty. In fact, the TFR of China was already as low as that of Korea or Taiwan at the end of the 1970s when the one-child policy was adopted. It is said that China's statistical system, especially that of birth registrations, deteriorated after the implementation of the one-child policy (Rindfuss and Choe 2015). Tian (2007) estimated that fertility dropped to the below-replacement level in 1992, which was approximately a decade later than in Korea and Taiwan. It seems that the good economy in 1985–95 boosted the birth rate despite the strict anti-natal intervention.

The most surprising demographic phenomenon after the turn of the century would be the emergence of extremely low fertility in the Republic of Korea and Taiwan. Table 6.1 compares the lowest and latest TFR values by country. Most English-speaking advanced countries and Northern/Western European countries sustained the level of 1.4. Although Denmark and Finland had TFR value less than 1.4, these countries quickly restored that line. Japan's lowest record of 1.26 was within the range of German-speaking countries. Southern European countries experienced lower fertility than German-speaking countries and Japan. It is understandable that some Eastern European countries recorded extremely low values of TFR because of the socioeconomic confusion during the transition to capitalism and a market economy. However, TFR values of Korea and Taiwan dropped to lower levels than those of Eastern European countries even though they did not experience such chaotic incidents.

The author's interpretation of ultra-low fertility in Korea and Taiwan is based on cultural determinism (Suzuki 2013). It is assumed that the degree of fertility decline in an advanced country depends on the gap between a rapidly changing socioeconomic system and a slowly changing family system. All advanced countries experienced more or less the same aspects of modernization, including economic development, the transition to a soft and service-oriented economy, a slow-down of the economic growth rate, globalization, increasing job insecurity, secularization, individualization, gender equalization, increasing marriage instability, increasing diversity of family types, and so forth. While the Northern/Western European family pattern was most adaptable to such changes, other societies experienced larger gaps because a family pattern could not change as rapidly as an educational or occupational system. It is interpreted that the larger the gap between the family and non-family systems, the lower the TFR goes down.

The Confucian societies had more difficulty in adapting to the post-modern socioeconomic conditions than the offspring of feudal societies in Japan or Europe. Features of the Confucian family that distinguish it from the feudal family included the emphasis on filial piety, low trust of non-family persons, rigid segregation of women, strict patrilineality and exogamy, and partible inheritance. It was inferred that the difference in gender equity between non-family and family systems is larger in Confucian societies than in Japan or other developed societies. The gap between the familial and non-familial institutions was not necessarily limited to gender relations but was also found in parent–child relations. The strong parent–child tie was supposed to promote a delay in the transition to adulthood and accelerated fertility decline. The notion that it is the mother's monopolistic role to care for a young child was assumed to restrict the supply and use of childcare services. Educational fever

Table 6.1 Recorded lowest and latest TFR

Region	Country	Lowest	(Year)	Latest	(Year)
English-speaking countries	United States	1.64	(2020)	1.66	(2021)
	New Zealand	1.61	(2020)	1.64	(2021)
	Australia	1.59	(2020)	1.70	(2021)
	Canada	1.41	(2020)	1.43	(2021)
Northern/Western Europe	France	1.66	(1994)	1.80	(2021)
	United Kingdom	1.53	(2021)	1.53	(2021)
	Sweden	1.50	(1999)	1.67	(2021)
	Netherlands	1.47	(1983)	1.62	(2021)
	Norway	1.41	(2022)	1.41	(2022)
	Denmark	1.38	(1983)	1.55	(2022)
	Finland	1.35	(2019)	1.48	(2021)
German-speaking countries	Switzerland	1.38	(2001)	1.51	(2021)
	Austria	1.33	(2001)	1.48	(2021)
	Germany	1.24	(1994)	1.58	(2021)
Southern Europe	Greece	1.23	(1999)	1.43	(2021)
	Portugal	1.21	(2013)	1.35	(2021)
	Italy	1.19	(1995)	1.25	(2021)
	Spain	1.16	(1998)	1.19	(2021)
Eastern Europe	Poland	1.22	(2003)	1.33	(2021)
	Slovak	1.19	(2002)	1.63	(2021)
	Russia	1.16	(1999)	1.49	(2021)
	Czech	1.13	(1999)	1.83	(2021)
	Bulgaria	1.09	(1997)	1.58	(2021)
Eastern Asia	Japan	1.26	(2005)	1.26	(2022)
	China	1.05	(2015)	1.30	(2020)
	Korea	0.78	(2022)	0.78	(2022)
	Taiwan	0.87	(2022)	0.87	(2022)

OECD Family Database, DGBAS

and a low evaluation of manual labor were thought to stimulate furious competition and raise educational cost.

The proposition "when gender equity rises to high levels in individual-oriented institutions while remaining low in family-oriented institution, fertility will fall to very low levels" (McDonald 2000) refers to the gap between the family and non-family systems in the gender issue. Table 6.2 shows the situation of gender equity in Eastern Asia in 2021–22. Since Taiwan is not a member of the United Nations, it appears neither in the *Human Development Report* by the United Nations Development Program (UNDP) nor in the *Gender Gap Report* by the World Economy

Forum (WEF). Instead, Gender Equality Committee of the Executive Yuan (2021, 2022) calculated Taiwan's score to be comparable with reports by UNDP and WFF. According to the result, Taiwan has achieved the highest gender equity in Eastern Asia.

The GII and GGI are based on such measures as women's political participation, education, labor force participation, promotion to senior officials and managers, professional and technical occupations, maternal mortality, adolescent fertility, and so on. Thus, it is assumed that both indices express the gender equity in non-family sectors. While Taiwan's gender equity in these sectors is impressive, there is some evidence that Taiwanese families sustain conservative attitudes. Figure 6.2 shows the change in sex ratio at birth (the number of male births per 100 female births) in Eastern Asia. Japan has never shown an abnormal ratio higher than 107 since 1980. On the other hand, Confucian societies with strong son preference started to show abnormally high ratios since the 1980s when prenatal sex identification such as villus examination and ultrasound method became available. The ratio used to be higher in Korea than in Taiwan until the mid-1990s. However, it seems that Korean parents have recognized that a son is not a source of old age security anymore, and the sex ratio dropped to the normal level of 106.1 in 2007.

The son preference is more robust in Taiwan, and the ratio has been fluctuating between 107 and 108 since 2010. Furthermore, Taiwan is the only country that participated in the National Transfer Account studies where the familial transfer played the most important role in the life of elderly persons (Lee et al. 2012). Given the very high gender equity in non-family sectors and low equity in the family sector, Taiwan is a typical case that McDonald's proposition applies. Such wider discrepancies between the familial and non-familial sectors suggests that fertility in Taiwan should be lower than in Korea. In fact, Taiwan showed lower TFR scores after 2006, even if fertility decline of Korea started earlier.

Since 2016, however, Korea has entered a new phase of fertility decline. Korea recorded TFR values as low as 0.98 in 2018 while Taiwan recorded 1.05. In 2022, the TFRs of Korea and Taiwan were 0.78 and 0.87, respectively. This requires another ad-hoc interpretation. A possible interpretation is that the disappointment toward the Moon Jae-In government (in office 2017–22) caused the recent acute decline of fertility. Young Koreans have experienced the difficulty in finding a stable job, working poor status, widening economic disparity, and generational succession of poverty. Neoliberal economic policies favoring chaebols adopted by the conservative governments of Lee Myung-Bak (in office 2008–13) and Park Geun-Hye (in office

Table 6.2 Gender equity in Eastern Asia (2021–22)

	Japan	Korea	China	Taiwan
Gender inequality index (2021)	0.083	0.067	0.192	0.036
Gender gap index (2022)	0.650	0.689	0.682	0.751

Gender Equality Committee of the Executive Yuan (2021, 2022)

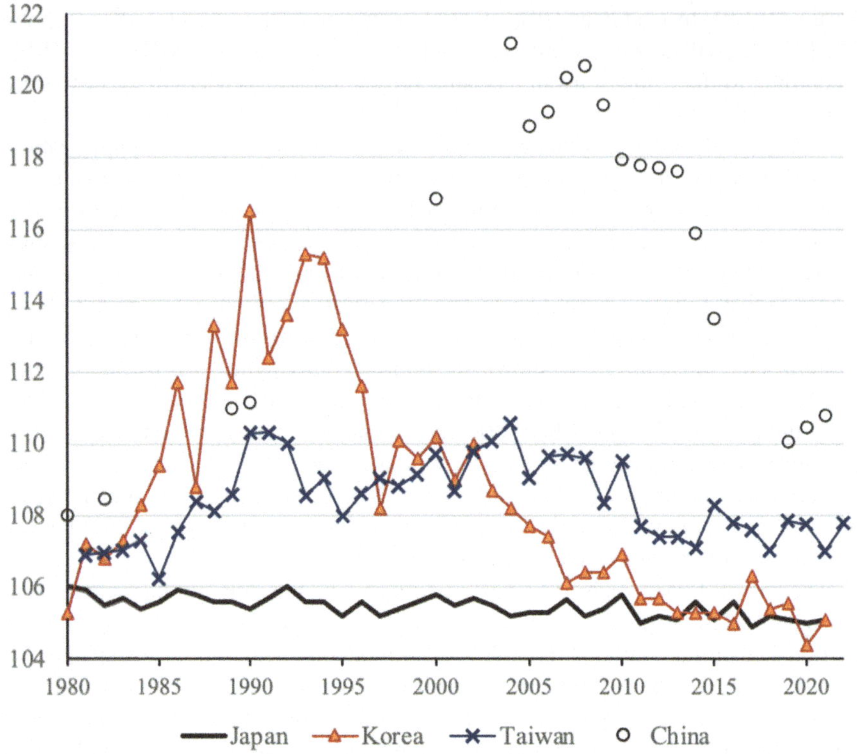

Fig. 6.2 Sex ratio at birth: 1980–2022 Japan: IPSS (2023a).Korea: Statistics Korea. Taiwan: DGBAS China: Nippo (2015), National Health Commission (2020, 2021, 2022).

2013–17) were blamed for such problems. Young people expected that the progressive government of Moon Jae-In, who took power in May 2017, would change the situation. However, Moon's economic policies such as raising the minimum wage, limiting the maximum hours that could be worked per week, or enforcing permanent employment, have resulted in slow economic growth, widening economic disparity, and a decline in job opportunities for young workers. Thus, it turned out that the change from a conservative to a progressive government could not improve the situation of young Koreans. Such a disappointment supposedly caused the recent acute decline in marriages and childbirths in Korea.

China's fertility transition is unclear due to the unreliable statistics. While the census result showed that the TFR value was already as low as 1.24 in 2000, the UNPD (2022) estimated that the value was 1.628. Some Chinese demographers have asserted that that census values (1.24 in 2000 and 1.19 in 2010) were realistic (Guo and Gu 2014). However, it is unlikely that such ultra-low fertility appeared when socioeconomic conditions was not stimulating low fertility. The Chinese economy was growing rapidly and the labor market condition for young workers was good. The college enrollment rate was lower than in advanced countries and the educational cost was not expensive. As shown in Fig. 6.2, the sex ratio at birth was extremely high during the 2000s and the early 2010s, suggesting that there was a strong son preference especially in rural areas. It is unbelievable that the combination of such socioeconomic conditions and the traditional attitude toward childbearing caused extremely low fertility recorded in the 2000 and 2010 censuses. It is also puzzling that the TFR rose from 1.047 in the 2015 micro-census to 1.30 in the 2020 census. Although Guo et al. (2018) asserted that 1.047 in 2015 was realistic, it is difficult to believe that the TFR of China was significantly lower than Korea (1.239) and Taiwan (1.188) at that time. It is unlikely again that the TFR rose in China while most countries experienced fertility decline due to the COVID-19. It is doubted if the National Statistics Bureau had a political intention to legitimize the abolishment of one-child policy.

China sustained the anti-natal policy longer than other parts of Eastern Asia. Table 6.3 compares changes in fertility policies in Japan, Korea, Taiwan, and China. As mentioned above, the Japanese government has never directly intervened into an anti-natal policy. The government decided to intervene after the "1.57 shock" in 1990 and published the government action program called "Angel Plan" in 1994. Under the condition of below-replacement fertility and in response to agreements made during the International Conference on Population and Development in Cairo in 1994, the Korean government decided in 1996 to abolish anti-natal population policy. In 2004, the government launched the Presidential Committee on Aging and Future Society to compile new population policies. After a discussion with representatives of business leaders, labor unions, activists, and feminists, the government action program "Saeromaji Plan 2010" was announced in 2006. The Taiwanese government abolished the numerical goal of population growth rate in 1992 and transitioned to the neutral policy on fertility. Taiwan's Ministry of the Interior planned to publish the *White Book of Population Policy* including pro-natal policy measures in 2005. However, feminist and ecologist groups protested the policy transition, emphasizing Taiwan's overpopulation. The *White Book of Population Policy* was finally published in March 2008.

China has sustained the anti-natal policy for three decades after below-replacement fertility was achieved in 1992. After the turn of the century, many experts have proposed banning the one-child policy. However, the former National Population and Family Planning Commission repeated that the policy will be firmly maintained during the Twelfth Guideline period (2011–15). In 2013, however, there was a relaxation of policy that a couple can have a second child if either spouse has

Table 6.3 Transitions in fertility policy

	Japan		Korea		Taiwan		China	
	Year	TFR	Year	TFR	Year	TFR	Year	TFR
Start of anti-natal policy	–	–	1961	5.81	1964	5.10	1979	2.67
End of anti-natal policy	–	–	1996	1.57	1992	1.73	–	–
Start of pro-natal policy	1994	1.50	2006	1.12	2008	1.05	–	–

no siblings. The relaxation proceeded gradually, and the restriction was abolished for the second child in 2016, and for the third child in 2021. Considering the length of below-replacement fertility and the levels reported in censuses, China's obsession to control fertility was pathological.

6.3 Mortality Decline and Epidemiological Transition

After WWII, there was a global trend of mortality decline owing to such innovations as sulfonamide, antibiotics, and high-yielding variant crops. As a result, the population of developing countries began to grow at extremely high growth rate of 2% or more. Eastern Asian countries other than Japan also had abnormally rapid population growth caused by the combination of declining mortality and remaining high fertility, which induced the governmental efforts of family planning programs. The emergence of below-replacement fertility in advanced countries after the 1970s implies that fertility declined too fast relative to mortality decline. Mortality decline has been slower than fertility decline also in Eastern Asia but has continued longer since the 1950s. Figure 6.3 shows the long-term trend of life expectancy at birth estimated by the UNPD (2022). Mortality decline in Eastern Asia was led by Japan, as in other socioeconomic and demographic changes. By 1980, Japan's life expectancy exceeded that of Sweden and became longest in the world. For the reasons of lowest mortality in Japan, Horiuchi (2010) pointed out the dietary habits with low fat and calorie, the cleanliness and health consciousness lasting from the Edo period, genetic factors including low ApoE4 frequency, relatively small income gap, and the cultural homogeneity with active vertical communication.

The epidemiological transition is the shift of the major causes of deaths from infectious to degenerative and man-made diseases. Omran (1971) classified Japan to the "accelerated model" characterized by the rapid decrease of infectious diseases promoted by the medical innovation and the public health intervention. Figure 6.4 shows the change in the share of category I labelled "infectious and parasitic diseases" in the total deaths. As in the early twentieth century, the vital statistics of Japan tabulated the number of deaths by cause defined in the ICD that has been maintained by the WHO since its foundation in 1948. The graph demonstrates how Japan's transition was drastic. While the category I accounted for 20.0% of all deaths in 1949, the share dropped to 6.1% in 1960 and 3.4% in 1970. This dramatic change was mainly caused by the decrease of tuberculosis. Table 6.4 shows the detail of deaths from infectious and parasitic diseases in 1950–70. The deaths from tuberculosis declined from 13.5% in 1950 to 2.2% in 1970. According to Ishi (2018), both the medical innovation and the public health intervention contributed to this rapid decline. Although streptomycin, the antibiotic especially effective for the tuberculosis, was expensive immediately after WWII, the domestic production began in 1949 and gradually reached to a massive number of patients. The tuberculin test and the BCG vaccination based on the TB Prevention Law in 1951 effectively prevented

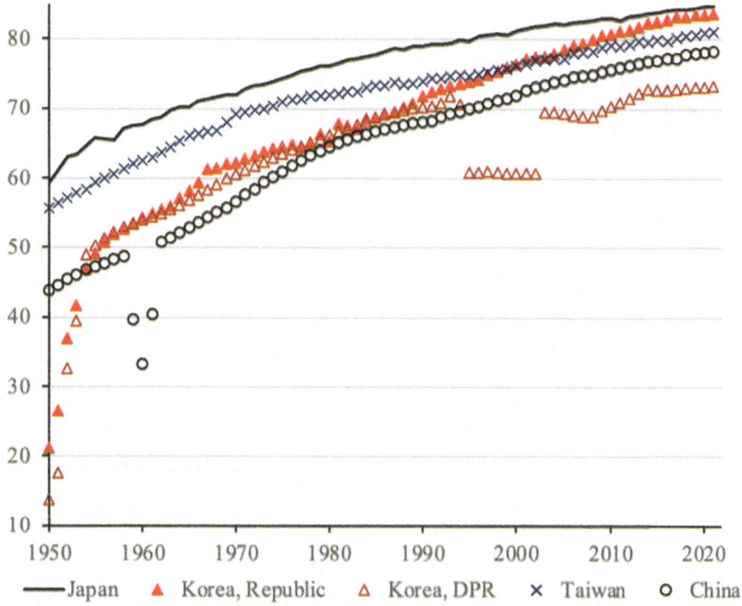

Fig. 6.3 Life expectancy at birth (both sexes): 1950–2021. UNPD (2022)

new incidents of the disease. As shown in Table 6.4, bacterial diseases other than the tuberculosis also declined significantly, suggesting the effect of antibiotics.

The Korean War broke out on 25 June 1950 by the invasion of DPR Korea. The armies of the US and the Republic of Korea were forced to the southern tip of the peninsula by the end of July. The tide turned when the UN forces landed on Incheon in September, and the DPR Korean army was pushed back to near the border with China. However, the frontline retreated again to the 38th parallel when China intervened in October. The battle lasted until the armistice was signed on 27 July 1953 and victimized a huge number of Koreans. Although the accurate number is unknown, it is estimated that two or three million Korean soldiers and civilians in the north and the south died. The estimated life expectancies at birth during the war are extremely low both in the Republic and the DPR Korea, as can be seen in Fig. 6.2. The mortality decline in the Republic of Korea was slow even after the war, and it took time to overcome infectious diseases. According to Kwon and Kim (2002), infectious and parasitic diseases still accounted for 11.4% of the total deaths in 1966 and decreased to 4.0% in 1974. It was in the late 1990s that the life expectancy of Korea caught up Taiwan. After the turn of the century, however, mortality has been significantly improved in the Republic of Korea and is approaching to the level of Japan.

In Eastern Asia, there were two great famines during the second half of the twentieth century. Both were man-made disasters caused by political failures. In November 1957, Mao Zedong declared in Moscow that China would catch up the United Kingdom within 15 years. The communist party approved the policy of "Great

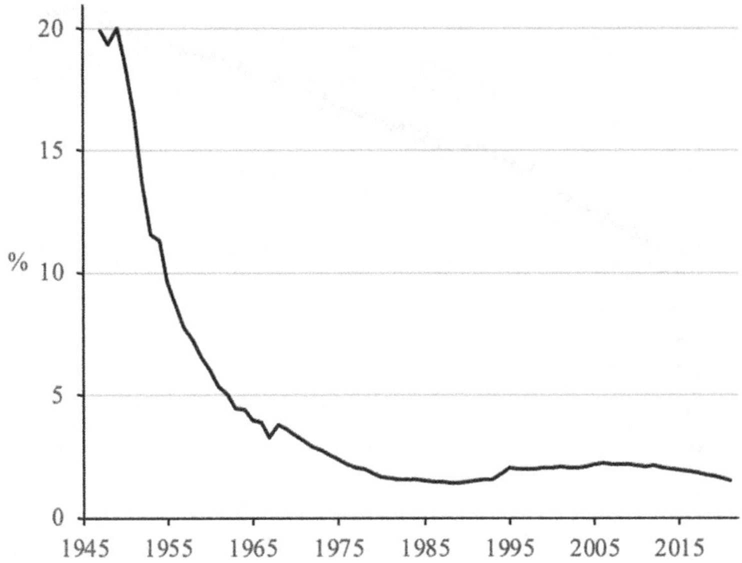

Fig. 6.4 Share of infectious and parasitic diseases in total deaths; Japan 1947–2021. Vital Statistics

Table 6.4 Deaths from infectious and parasitic disease: Japan 1950–70

	No. of deaths			Percent		
	1950	1960	1970	1950	1960	1970
Total	904,876	706,599	712,962	100.0	100.0	100.0
Infectious and parasitic diseases	167,978	42,766	24,163	18.6	6.1	3.4
Tuberculosis	121,769	31,959	15,899	13.5	4.5	2.2
Intestinal infectious diseases	12,974	2,227	4,798	1.4	0.3	0.7
Other bacterial diseases	13,772	2,341	961	1.5	0.3	0.1
Siphilis and other STDs	5,176	2,068	662	0.6	0.3	0.1
Viral diseases	7,794	3,271	779	0.9	0.5	0.1
Other infectious and parasitic diseases	6,493	900	1,064	0.7	0.1	0.1

Vital Statistics of Japan

Leap Forward" proposed by Mao in 1958. Because Mao ordered to industrialize rural China without relying on foreign investments or aids, the advice of Russian experts was ignored, and domestic intellectuals were expelled as right wings. Although a huge number of backyard furnaces were built and farmers were mobilized to satisfy the unrealistic production goal set by Mao, outputs were mostly useless for any industrial purpose. Because male farmers were mobilized to the hard work in iron manufacture, agricultural labors were left to unexperienced women. The agricultural production declined sharply because of inadequate planting and insufficient weeding. On the other hand, a plenty of foods were supplied to workers mobilized to the heavy industry, and even seeds for the next year were eaten. Because imported foods were sent to cities, a huge number of excess deaths occurred mainly in rural areas. According to Dikoetter (2010), demographers' estimates of excess deaths in 1958–62 ranged between 23 and 36 million. He criticized that these were underestimations and asserted that there were more than 45 million excess deaths. As can be seen in Fig. 6.3, the UNPD (2022) considered that mortality of China was abnormally high in 1959–61. Table 6.5 shows the number of total deaths and CDRs in China taken from the UNPD (2022). In this table, the "normal" CDR is simply the mean of CDRs in 1958 and 1962. If we assume that the excess death rate is the difference from the normal rate, the number of excess deaths can be obtained by applying the percentage of excess deaths to the total number of deaths. The calculated number of excess deaths for three years is 17 million, which is smaller than the lowest estimate in the past.

By the 1990s, the global network of communication and transportation was well established and international organizations such as the World Food Programme acted to monitor the food shortage and coordinate international aids. As a result, even during the notorious "apocalypse drought" that hit Southern Africa in 1991–92, there was no prominent mortality rise (Messer and DeRose 1998). Thus, the great famine in the DPR Korea was a shocking incident because it occurred in the era supposed to be free from famine. The economy of the DPR Korea already stagnated in the early 1980s due to the limit of a planned economy. The food shortage was seriously aggravated when the supply from the Soviet Union was disrupted in 1991. The floods in July and August of 1995 destroyed the barely operated rationing system and the DPRK government had to request for international aids. According to Moon (2011), estimates of excess deaths ranged in a wide range between 270,000 and 3,500,000. His own demographic analysis showed that the death toll in 1994–2000 was about 336,000. Table 6.6 shows the size of excess deaths implied in the UNPD (2022). The UN demographers assumed that abnormally high mortality lasted for eight years until 2002 and that there were 661,000 excess deaths.

Table 6.5 Excess deaths in China 1959–62

	Total deaths (thousand)	CDR (‰)	Excess death (%)	Excess death (thousand)
1958	11,733	18.2		
1959	16,175	24.8	31.8	5,141
1960	19,613	30.0	43.5	8,536
1961	14,553	22.2	23.8	3,457
1962	10,426	15.6		
	Normal	16.9	Total	17,134

Based on UNPD (2022)

Table 6.6 Excess deaths in DPR Korea 1995–2002

	Total deaths (thousand)	CDR (‰)	Excess death (%)	Excess death (thousand)
1994	130	5.9		
1995	221	9.9	30.0	66
1996	226	10.0	31.0	70
1997	231	10.1	31.9	74
1998	238	10.3	33.2	79
1999	245	10.6	34.8	85
2000	251	10.8	35.9	90
2001	258	11.0	37.2	96
2002	263	11.1	38.1	100
2003	188	7.9		
	Normal	6.9	Total	661

Based on UNPD (2022)

6.4 Population Growth Rate and Age Structure

Figure 6.5 shows annual population growth rates of Japan, the Republic of Korea, Taiwan, and China based on official statistics. The growth rate of Taiwan was high in 1986 because Kinmen islands were included into the total population from this year. The growth rate of Japan was high in 1972 because Okinawa was returned to Japan in this year. The population of China decreased due to the great famine in 1959–60. Because immigration has been strictly limited in Eastern Asia, the population growth has been dominated by the trend of fertility and mortality. The growth rate was very high in the 1950s and 1960s owing to the combination of high fertility and declining mortality. While Japan never experienced the rate of 2% or higher except for 1972, extremely high growth rates continued in Korea, Taiwan, and China until the 1970s. Although Japan reached the below-replacement level as early as in 1957, the TFR rose again and exceeded 2.1 in the early 1970s. While fertility never regained the replacement level after 1974, the population growth continued for three decades because of the momentum of the past. Japan experienced temporary population decline in 2005 and 2008 before the continuous decline started in 2010. Thus, it took about 30 years for the population momentum of Japan to extinguish. In China, it also took three decades from the arrival at the below-replacement level in 1992 to the beginning of population decline in 2022. On the other hand, the population growth under below-replacement fertility lasted for more than 35 years in Korea and Taiwan probably due to the larger momentum than Japan and China.

An increasing population is young because larger weights are assigned to younger cohorts. Thus, populations in Eastern Asia were young owing to the high growth rates in the 1950s and 1960s. It was supposed that a rapid population growth obstacles economic development through a high child dependency ratio and a diversion of governmental investment (Kelley 1988). Population aging begins as soon as fertility

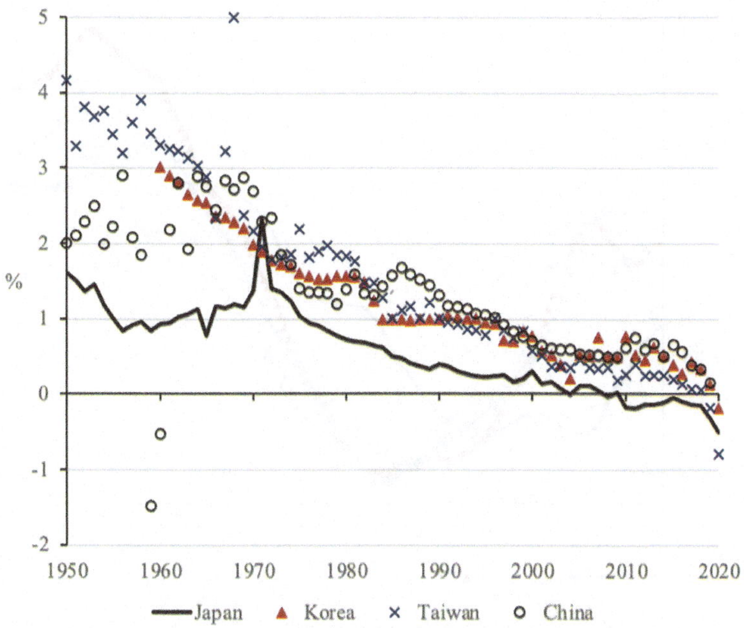

Fig. 6.5 Population growth rate: 1950–2020. Japan: IPSS (2023a). Korea: Statistics Korea, Taiwan: Ministry of the Interior (2009), DGBAG. China: Wakabayashi and Nie (2012), National Bureau of Statistics

starts to decline. However, because the percentage of working-age population (15–64 years old) grows more quickly than that of the elderly population (65 years and older), the total dependency ratio (i.e., children and elderly populations divided by the working-age population) declines for several decades. This decrease in net consumers relative to net producers is referred to as a "demographic bonus" or "demographic gift" (Mason 2001), or a "demographic window of opportunity" (Lee 2007).

Figure 6.6 shows the long-term trend of total dependency ratios in Japan, Korea, Taiwan, and China based on the medium variant of the UNPD (2022). Because fertility decline reduces the size of the child population, the total dependency ratio declines at the early stage of fertility transition. However, the ratio finally starts to rise as the effect of population aging becomes dominant. The turning point was the late 1990s for Japan and 2010–15 for Korea, Taiwan, and China. Considering that Japan experienced rapid economic growth in the 1950s and 1960s and that Korea and Taiwan experienced it in the 1970s and 1980s, it can be said that these three economies made full use of the demographic gift. However, the demographic gift of China ended before catching up developed countries. China's failure to synchronize its economy and demography is a recurrence of that of the nineteenth century. The Chinese population started growing too early to learn how to break through the ceiling of an agricultural society. In the twentieth century, the Chinese economy stagnated because of the political confusion, while fertility decline proceeded as fast as Korea

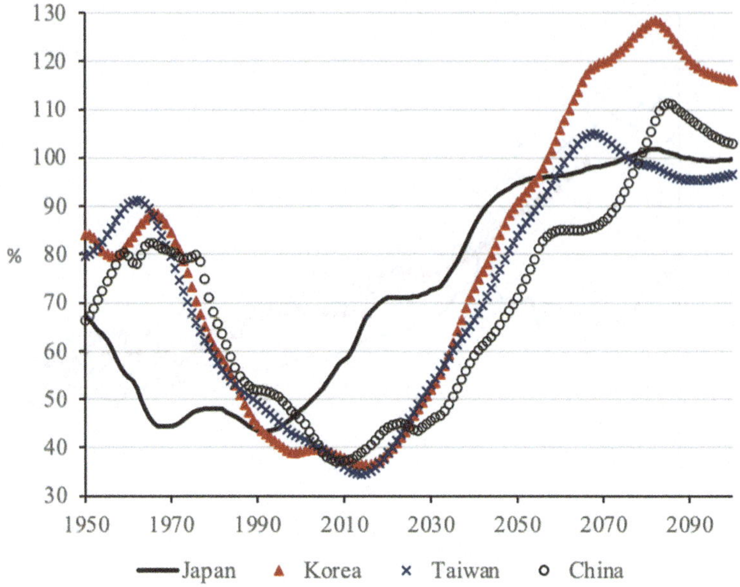

Fig. 6.6 Total dependency ratio: 1950–2100. UNPD (2022)

and Taiwan. This dissonance between economy and demography is what caused the problem of early aging relative to economic development called "wei-fu xian-lao" in Chinese.

According to the medium variant of the UNPD (2022), the total dependency ratio of Korea will exceed that of Japan in 2055. Taiwan will also overtake Japan in 2060, and China in 2080. This implies that the population aging in Korea, Taiwan, and China will be more drastic because fertility will stay at a lower level than Japan. Table 6.7 shows TFRs, life expectancies at birth, total dependency ratios, and percentages of the elderly aged 65 years and older in specific years assumed and projected by the UNPD. Projection results show that the population aging in Korea will be most serious because of the lowest fertility level and the mortality level as low as Japan. It is projected that Korea will overtake Japan in 2046 in terms of the percentage of the elderly aged 65 years and older. Taiwan also will have an older population than Japan in 2059, and China will do so in 2077. Thus, Eastern Asia will be the most aged region worldwide in the second half of this century.

The tempo of population aging in Eastern Asia can be faster than the projection by the UNPD. Table 6.8 shows the latest projection results by the IPSS (2023b), Statistics Korea (2021), and the National Development Council (2022). No governmental official population projection was found for China. Agencies in Japan, Korea, and Taiwan are apparently more pessimistic about the recovery of fertility than the UNPD. It is projected that the total dependency ratio of Korea will overtake Japan in 2045 and the percentage of the elderly in 2043. The population aging of Korea is much faster than Japan because the Statistic Korea assumed not only lower fertility

Table 6.7 Age structure projected by united nations population division

	Japan	Korea	Taiwan	China
Total fertility rate				
2050	1.47	1.17	1.41	1.39
Life expectancy at birth (both sexes)				
2050	88.3	87.4	85.6	83.8
Total dependency ratio				
2050	94.7	90.8	83.9	71.1
2055	95.9	96.3	90.2	81.1
2060	96.3	105.8	97.5	85.0
2080	101.4	127.5	98.5	103.2
Percentage of elderly aged 65 and older				
2046	36.9	37.3	33.0	28.2
2050	37.5	39.4	35.3	30.1
2059	38.2	43.0	38.4	35.3
2077	38.8	47.1	38.6	39.1

Medium variant of UNPD (2022)

but also lower mortality than Japan. According to a comparison between official projections, Taiwan will overtake Japan in 2053 in terms of the total dependency ratio and in 2049 in terms of the percentage of the elderly. In any case, it is certain that Korea and Taiwan will be more aged society than Japan in the middle of this century.

6.5 Marriage and Divorce

Figure 6.7 shows the long-term trend of the percentage remaining single for women aged 25–29 years old in Japan, Korea, and Taiwan. The pattern of early and universal marriage was apparent in Eastern Asia in the 1950s and 1960s. In Korea and Taiwan, more than 90% of women in this age group were already married. Although Japan showed a temporary change to opposite in the 1960s, the secular trend was the delay of marriage due to urbanization, the popularization of higher education, and the increasing female labor participation (Smith 1980).

It was around 2005 that the percentage of single women in Korea and Taiwan exceeded that of Japan. Such a drastic decline in nuptiality was the direct cause of lowest-low fertility in these societies. Because extramarital births are rare in Eastern Asia, a decline in marriage rate immediately results in fertility decline. While other elements of second demographic transition have already taken place in Eastern Asian advanced countries, only the increase in extramarital births cannot be observed in this region (Lesthaeghe 2010). Although it is difficult to explain the reason, one

Table 6.8 Age structure projected in official projections

	Japan	Korea	Taiwan
Total fertility rate			
2050	1.35	1.21	1.20
Life expectancy at birth (both sexes)			
2050	87.5	88.9	85.5
Total dependency ratio			
2045	86.5	86.9	77.6
2050	89.0	95.8	87.5
2053	89.5	97.8	90.7
Percent of elderly aged 65 and older			
2043	35.8	36.1	32.9
2049	37.0	39.8	37.0
2050	37.1	40.1	37.5

Japan: IPSS (2023b)
Korea: Statistics Korea (2021)
Taiwan: National Development Council (2022)

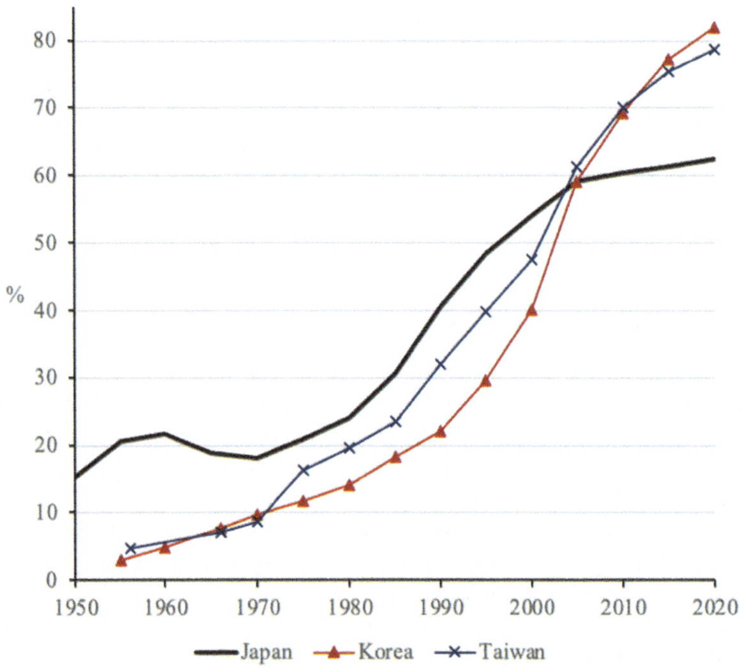

Fig. 6.7 Percentage of single women aged 25–29 years old: 1950–2020. Censuses, DGBAS

important point should be the role of secularization. Inglehart extracted two axes of modernization, namely from "traditional" to "secular-rational" and from "survival" to "self-expression" (Inglehart and Welzel 2008). In societies of European origin, the prevalence of oral contraceptives and the increase of premarital intercourse were driven by the liberation from the religious authority of church. On the contrary, Eastern Asian countries were originally quite secular societies and there was no religious authority comparable to the Christian church after WWII. While the secularization was the catalyst of reproductive revolution in advanced countries in Europe and Northern America, the disconnection of childbirth from marriage did not happen in Eastern Asia because there was little change in religious belief.

Figure 6.8 compares the percentage of divorced women in the age group of 45–49 years old. As in the percentage of single women, the lead of Japan lasted long after WWII. It was in the 1990s that divorces increased rapidly in Taiwan and exceeded Japan. Korea also caught up Japan around 2010. The drastic changes in marriage, divorce, and childbirth in Confucian societies since the 1990s can be attributed to the expanding discrepancy between the rapidly changing non-family system and the slowly changing family system. Especially in Taiwan, the gender equity in education, occupation, and politics was smoothly improved and achieved the highest level in Eastern Asia. As a result, the changes in marriage and divorce were faster than in Korea. On the other hand, the labor market condition for young workers was worse in Korea where the 1997 financial crisis happened. Thus, the percentage of single women quickly caught up Taiwan and the TFR reached the lowest-low level earlier than Taiwan.

It is difficult to compile a long-term series of marital states in China. According to the 2010 census report, only 21.6% of women aged 25–29 years old remained single, which was lower than Japan in 1980. The percentage of divorced women (2.1%) was also lower than Japan in the same year. Thus, there was no drastic change in marriage and divorce as in other advanced societies in Eastern Asia. As discussed above, the TFR of 1.19 in the 2010 census is difficult to believe because it was recorded when the economic growth rate was high and the educational cost was low. The absence of prominent nuptiality decline can be added the reasons why the TFR of China is unreliable.

6.6 Contemporary Household Changes

Figure 6.9 shows the long-term trend of the mean household size in Eastern Asia. In Japan, the basic types of households have been "ordinary households" and "quasi households," the latter including boardinghouses, company's dormitories, student dormitories, hospitals, social institutions, etc. Since 1970, a new distinction between "private households" and "institutional households" has been added and persons in boardinghouses and company's dormitories were included to the former type. Although the new definition of private household is more widely used today, the mean size of ordinary households was used in the graph to show a long-term trend.

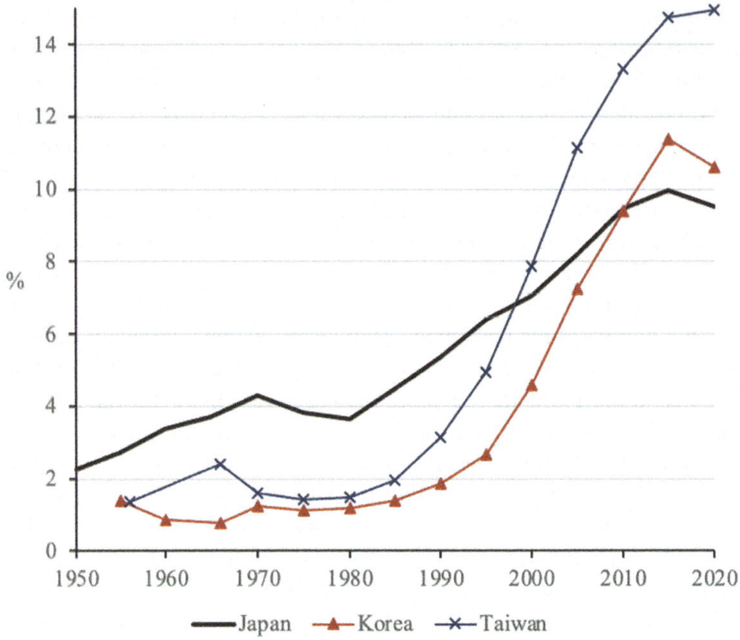

Fig. 6.8 Percentage of divorced women aged 45–49 years old: 1950–2020. Censuses, DGBAS

In 1970, the mean sizes of ordinary households and private households were 3.69 and 3.41, respectively. However, the difference has been less than 0.1% point since 1985.

The decline in household size can be attributed to urbanization in addition to fertility decline. In Japan, there was a massive flow from rural to urban areas during the rapid economic growth in the 1950s and 1960s. While those in-migrants formed nuclear family households in cities, traditional stem family households decreased because even designated heirs left home for more beneficial occupations than agriculture. It is supposed that such a shift from complex to simple household types occurred in other parts of Eastern Asia especially when the economy grew fast and migration to cities was substantial. The data for China in 1953 and 1964, taken from the census report by the Office of the Leading Group of the State Council (2021) are suspicious. It is unlikely that the average household size in 1953 was smaller than Japan and that there was no significant change for 30 years.

Figure 6.10 shows the change in percentage of one-person households. While the one-person households accounted for less than 10% until the 1960s, currently the household type has the largest share in Eastern Asia. The data for Taiwan are less trustworthy and it is doubted that if accelerations and decelerations shown in the graph really happened. In any case, unlike in nuptiality and fertility changes, Korea and Taiwan have not yet overtaken Japan. In 2020, 36.8% of ordinary households were one-person households in Japan. This means that 36.8% of ordinary household

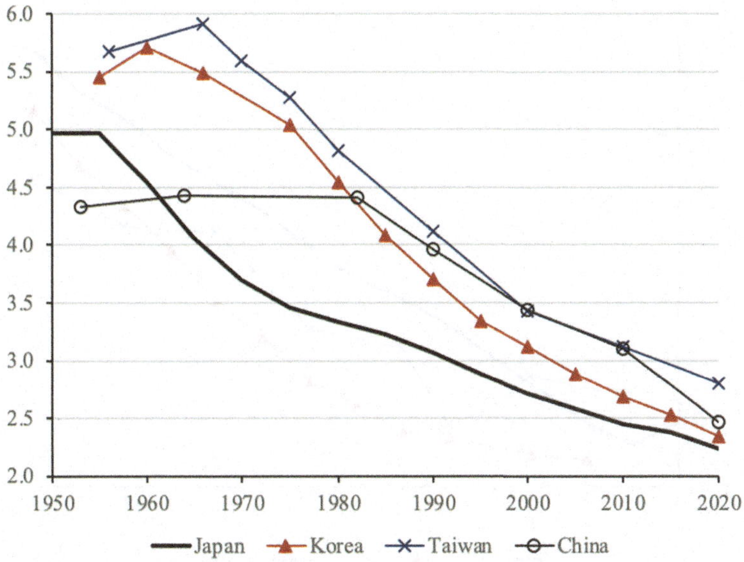

Fig. 6.9 Mean household size: 1950–2020. Censuses

heads were living alone. Because the mean size of ordinary household was 2.23, the percentage of ordinary household members living alone was 36.8/2.23 = 16.4% in 2020. This percentage, or the propensity to live alone, was 13.5% in Korea and 9.2% in Taiwan in 2020. In the 2010 census of China, the mean size of family households was 3.10 and one-person households accounted for 14.5%. Then, the propensity to live alone was 4.7%. The comparison of marriage, divorce, and household suggests that China is still behind Japan, Korea, and Taiwan in most demographic changes.

6.7 Urbanization in Contemporary Period

Figure 6.11 shows changes in Gini coefficients of population distribution since 1945. Japan's Gini coefficient rose sharply during the period of economic growth in the 1950s and 1960s. The change was slow until the early 1990s but slightly accelerated after the land prices fell in the late 1990s. In fact, the net in-migration rate of the Tokyo metropolitan area turned from negative to positive after 1995. The collective share of Tokyo, Saitama, Chiba, and Kanagawa prefectures in the total population increased from 13.0% in 1945 to 29.3% in 2020. Thus, more than one-quarter of the Japanese population live in Tokyo and neighboring prefectures that account for only 3.5% of the land area.

The Republic of Korea experienced the largest change in population distribution in Eastern Asia. As in Japan, the peak of urbanization coincided with the peak of economic growth in the 1970s and 1980s. According to the UNPD (2018), the

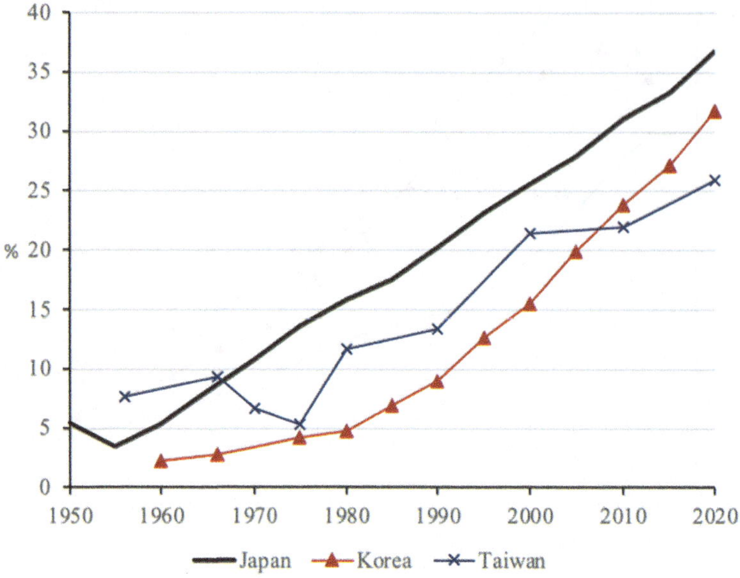

Fig. 6.10 Percentage of one-person households: 1950–2020. Censuses

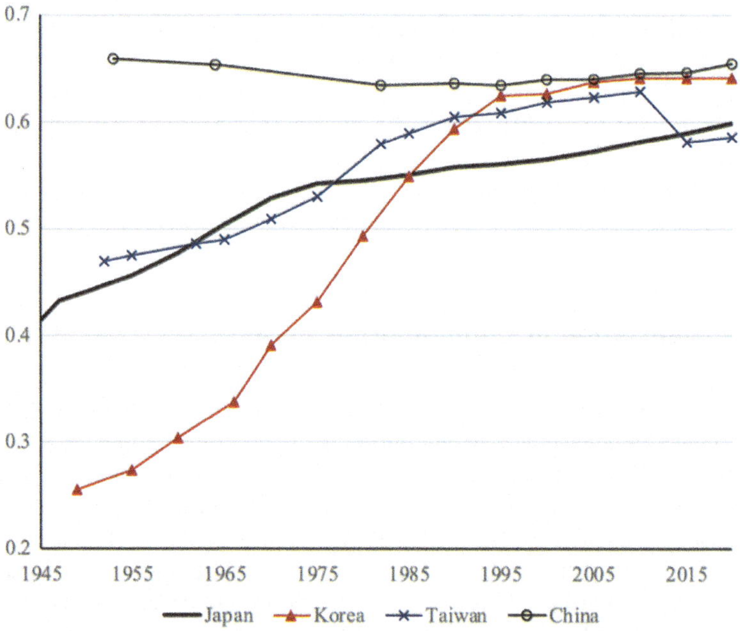

Fig. 6.11 Gini coefficient of population distribution in Eastern Asia: 1945–2020. Censuses, DGBAS

percentage of the urban population in the Republic of Korea increased from 21.4% in 1950 to 86.4% in 2015, which was the largest change in Eastern Asia. The share of the capital area, including Seoul city, Incheon city, and Gyeonggi province grew from 20.7% in 1949 to 50.5% in 2020. Thus, one half of the total population inhabits the capital area, which accounts for only 11.8% of the land area.

The change in population distribution in Taiwan has been milder than that of Korea. As a result, the concentration of population is not as extreme as it is in Korea. In 2020, 45.6% of total population inhabited the northern part of Taiwan, consisting of Taipei city, New Taipei city, Keeling city, Taoyuan city, Hsinchu city, Hsinchu county, and Yilan county, which accounts for 20.4% of land area. In December 2010, the former Taipei county changed to be New Taipei city. In addition, the former Taichung, Tainan and Kaohsiung counties merged with Taichung, Tainan, and Kaohsiung cities. Because of this administrative reformation, the Gini coefficient since 2015 is not comparable with the past.

The Gini coefficient of China based on provincial populations cannot capture the trend of urbanization. According to the UNPD (2018), the percentage of urban population increased from 11.8% in 1950 to 55.5% in 2015. The change is more acute than in Japan (from 53.4 to 91.4%) but milder than in Taiwan (from 26.8 to 76.9%). The land area and population of China are so large that concentration in only one center is impossible. According to the UNPD (2018), six Chinese cities (Shanghai, Beijing, Chongqing, Tianjin, Guangzhou, and Shenzhen) ranked in the 30 largest urban agglomerations in 2015.

According to the UNPD (2018), the Kanto Major Metropolitan Area (MMA), including Tokyo, Yokohama, Kawasaki, Chiba, Saitama, and other cities, with a population of 37 million, is the largest urban agglomeration in the world. This position is not necessarily new. As already noted, Edo's population had exceeded one million in the eighteenth century, which was probably the largest worldwide at that time. The concentration in Edo was provoked by a policy of the Tokugawa shogunate to force feudal lords to travel to Edo regularly. Since Osaka was economically more developed than Edo, there were two centers in Tokugawa Japan. The development of major cities, including Tokyo and Osaka after the Meiji Restoration has been already discussed in the previous chapter.

Table 6.9 shows populations of two largest cities and the ratio of the largest to the second largest city. In this table, "Tokyo" refers to the population of Kanto MMA, and "Osaka" refers to that of Kinki MMA including Osaka, Kyoto, Kobe, and other cities. The ratio of Kanto to Kinki fluctuated around 1.6 until 1975, implying that the populations of two metropolitan areas grew at a similar rate. However, the ratio increased as centralization in the Tokyo area became prominent. In the 1980 census, the population of Yokohama city exceeded Osaka city.

Figure 6.12 compares the annual population growth rates of four capital cities in Eastern Asia. Here, "Tokyo" refers to the population of 23 special wards. The former Tokyo City was abolished in 1943 and 35 wards were reorganized into 23 special wards in 1947. Although the population of central Tokyo grew rapidly in the 1950s, the growth turned negative in the late 1960s due to suburbanization. Because of the high land price in downtown areas, married couples out-migrated to neighboring

Table 6.9 Population (1000) of two largest cities and ratio of the largest to the second city

Year	Japan		Ratio	Korea			Taiwan			China		
	Tokyo	Osaka		Seoul	Busan	Ratio	Taipei	Kaohsiung	Ratio	Shanghai	Beijing	Ratio
1950	11,275	7,005	1.61	1,021	948	1.08	551	287	1.92			
1955	13,713	8,623	1.59	1,553	1,046	1.48	696	348	2.00			
1960	16,679	10,615	1.57	2,361	1,154	2.05	880	423	2.08	6,865	3,900	1.76
1965	20,284	13,065	1.55	3,452	1,365	2.53	1,113	514	2.17	6,428	4,582	1.40
1970	23,298	15,272	1.53	5,312	1,813	2.93	1,406	624	2.25	6,052	4,426	1.37
1975	26,615	16,298	1.63	6,808	2,418	2.82	1,778	757	2.35	5,658	4,828	1.17
1980	28,549	17,028	1.68	8,244	2,906	2.84	2,247	920	2.44	5,928	5,366	1.10
1985	30,304	17,583	1.72	9,334	3,473	2.69	2,497	1,117	2.24	7,096	6,017	1.18
1990	32,530	18,389	1.77	10,518	3,778	2.78	2,737	1,356	2.02	8,606	6,788	1.27
1995	33,587	18,940	1.77	10,256	3,813	2.69	2,704	1,431	1.89	11,072	8,355	1.33
2000	34,450	18,660	1.85	9,879	3,673	2.69	2,642	1,488	1.78	14,247	10,285	1.39
2005	35,622	18,762	1.90	9,822	3,533	2.78	2,636	1,504	1.75	17,056	12,991	1.31
2010	36,860	19,313	1.91	9,796	3,422	2.86	2,654	1,514	1.75	20,314	16,441	1.24
2015	37,256	19,305	1.93	9,897	3,446	2.87	2,686	1,525	1.76	23,482	18,421	1.27

UNPD (2018)

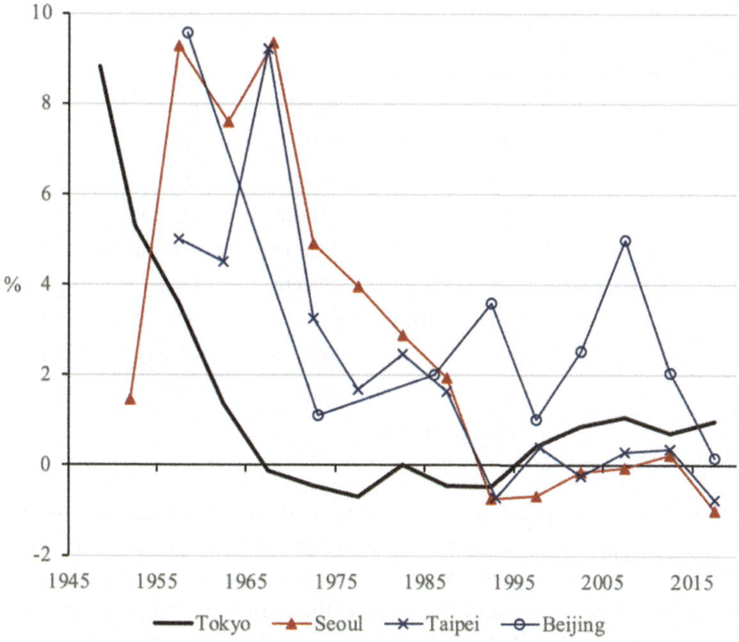

Fig. 6.12 Annual growth rate of capital city. Censuses, DGBAS, China Statistical Yearbook

prefectures to find reasonable housing. The population decline continued to the early 1990s but turned positive again as the land price fell. Koike (2017) pointed out that the declining position of Kinki MMA and in-migration from western Japan contributed to the recent re-centralization of Tokyo.

While MMAs are defined by the Statistics Office in Japan, the Statistics Korea does not define such areas. Thus, populations of Seoul and Busan in Table 6.9 are those of cities not including suburban areas. According to the estimation by the UNPD (2018), Seoul's population in 1950 was approximately the same as Busan's because of the Korean War. After the war, however, the growth rate of Seoul exceeded that of Busan. The ratio of the population of Seoul to Busan reached 2.87 in 2015, which was significantly higher than that of Tokyo, Taipei, or Shanghai. Figure 6.12 shows that the growth rate of Seoul was significantly higher than that of Taipei, until both cities confronted suburbanization at the same time in the early 1990s.

The cause of compressed urbanization and extreme concentration in the Republic of Korea was the massive outflow of the surplus population from rural areas that started in the Japanese period. Even after the Korean War, the agricultural development was too slow to keep the growing population within villages. Because the government emphasized heavy industrial plants located in major cities such as Ulsan, Masan, Pohang, and Changwon, these cities absorbed a large portion of population from rural areas. At the same time, satellite cities such as Suwon, Incheon, Bucheon, Seongnam, and Anyang developed around Seoul (Chang 2002).

Table 6.10 Percentage of population aged 65 years and over by urban and rural areas

Area			65+ (%)
Japan (2020)	Urban	(Shi)	27.6
	Rural	(Gun)	33.4
Korea (2020)	City	(Dong)	14.4
	Town	(Eub)	16.8
	Village	(Myeon)	29.9
Taiwan (2020)	Metropolis	(Zhixiaoshi)	15.7
	County	(Xian)	17.0
China (2020)	City	(Chengshi)	10.8
	Town	(Zhen)	11.8
	Village	(Xiangcun)	17.7

Census, DGBAS

Because in-migrants to urban areas tend to be young, the difference in age structure between urban and rural increases in the process of urbanization. Table 6.10 shows how differential magnitudes of urbanization resulted in different age structures. Naturally, urban age structure is younger than rural age structure in all countries. However, it is only in the Republic of Korea that the urban–rural difference in the percentage of elderly population exceeds 10% points. In fact, rural Korea was already more aged than urban Japan in 2020. It is supposed that the extremely acute urbanization of Korea has caused various problems. Among them, the worsened welfare of elderly Koreans can partly be attributed to the urbanization and underdevelopment of rural areas. Korea has been notorious for the highest poverty and suicide rates of the elderly population among OECD member countries (Onishi 2014; Suzuki 2021). It is plausible that financial and emotional supports from children to old parents are smaller in Korea than in other Asian countries because many parents are left behind in rural areas, and children in urban areas are too busy to sustain their own lives.

The issue of capital relocation was seriously discussed in the Republic of Korea in the 1990s owing to the extreme concentration of population, wealth, and political-economic functions. The former president Roh Moo-Hyun won in the 2002 election by promising to relocate the capital to the Chungcheong area. When the Constitutional Court declared in October 2004 that the capital relocation required a constitutional amendment, Roh's plan was amended to relocate ministries other than justice, foreign affairs, and unification to a newly founded "administrative capital." The former Yeongi County was renamed Sejong city and decided as the location of the administrative capital.

When Lee Myung-Bak succeeded to the presidency in 2008, he opposed the relocation of governmental functions. He attempted to transform Sejong city into an industrial, science, and education hub. However, his proposal was rejected in the parliament, and the relocation was executed according to the original plan. The relocation began in 2012; the Prime Minister's Office, nine ministries (including Ministry of Economy and Finance; Ministry of Land, Transport and Maritime Affairs;

Ministry for Food, Agriculture, Forestry and Fisheries; Ministry of Culture, Sports and Tourism; Ministry of Health and Welfare), and related agencies moved to Sejong city. On the other hand, organizations that remained in Seoul included the Presidential Office; Parliament; Supreme Court; Ministry of Foreign Affairs; Ministry of Justice; Ministry of Defense; Ministry of Unification; and the Ministry of Gender Equality and Family.

The projected demographic impact of the capital relocation is not very impressive. Table 6.11 shows the population share by region, based on the regional population projection by the Statistics Korea (2022). While the population share of Seoul City is expected to continue decreasing, those of Incheon city and Gyonggi province will increase. It is projected that the share of capital area will grow from 50.2% in 2020 to 53.0% in 2050. On the other hand, the share of Sejong city and Chungcheong province will increase by only 0.7% point each, implying that the Gini coefficient will continue to increase.

The urbanization of Taiwan was not as drastic as that of Korea. The increase in the Gini coefficient shown in Fig. 6.11 was as moderate as that of Japan. The agriculture of Taiwan sustained its competitiveness and supported economic growth by exporting sugar, rice, tea, and fruits. When the US required Taiwan to open its market, numerous small and medium-sized industries were founded in rural areas. Because such rural factories produced fabric, plastic, or electric products and absorbed the surplus population, the domestic migration from rural to urban areas was moderate (Ishida 2005). Unlike in Korea where a small number of conglomerates and gigantic chaebols governed the economy, many small and medium-sized enterprises grew in Taiwan by exporting products to the US and Japan. With the government's protection, these enterprises could avoid the influence of international conglomerates (Vogel 1991). Such different modes of industrialization resulted in the different tempo of urbanization in Korea and Taiwan.

Table 6.11 Projected regional population of Korea (%)

	Seoul	Incheon	Gyonggi	Capital area (subtotal)	Sejong	Chungcheon	Other
2000	21.4	5.4	19.5	46.3		7.2	46.6
2005	20.8	5.4	22.1	48.2		7.1	44.7
2010	20.4	5.5	23.4	49.3		7.3	43.4
2015	19.5	5.7	24.4	49.5	0.4	7.2	42.9
2020	18.6	5.7	26.0	50.2	0.7	7.3	41.8
2025	17.9	5.8	27.2	50.9	0.8	7.5	40.8
2030	17.5	5.8	28.2	51.4	1.0	7.6	40.0
2035	17.2	5.8	28.9	51.9	1.1	7.7	39.3
2040	17.0	5.9	29.5	52.4	1.2	7.8	38.7
2045	16.9	5.9	29.9	52.7	1.2	7.9	38.1
2050	16.7	5.9	30.3	53.0	1.3	8.0	37.7

Statistics Korea (2022)

While the population growth rate of Taipei was lower than that of Seoul in 1955–90, both capitals experienced a population decrease at the same time because of suburbanization. This can be added to the list of synchronized developments in Korea and Taiwan that includes the timing of democratization, the arrival at the replacement level of fertility, the emergence of the lowest level of fertility worldwide, and the expected position of the most aged population in the world. Japanese rule until 1945 is assumed to have determined these time schedules of development because Korea and Taiwan have been "offspring of the Empire," following the Japan model and utilizing the legacy of the colonial period (Eckert 1991). In addition, Korea and Taiwan were equally important as divided nations confronting the socialist world. Thus, the US and Japan eagerly supported these two economies.

The Gini coefficient of China in Fig. 6.11 declined until 1990 because of the migration to frontiers such as Shinjang, Tibet, Inner Mongolia, and Hainan. After 1990, however, the accelerated population growth in already populous provinces such as Guangdong and Zhejiang reversed the trend, and the population distribution became more uneven. In any case, the change in the Gini coefficient based on provincial population does not capture the trend of urbanization because it does not reflect the changing distribution within a province.

The restriction of residential movement through the "hukou" (household registration) system started in 1958. In 1964, the government ordered further strict restrictions of flows into cities. The legal discrimination between urban and rural residents has been sustained until today (Wakabayashi 1996). During the Cultural Revolution (1966–76), many urban intellectuals were sent to rural labor camps, which depressed urbanization (Ping 2009). As a result, the percentage of the urban population estimated by the UNPD (2018) declined from 18.1% in 1965 to 17.4% in 1975.

As the economic reform led by Deng Xiaoping started at the end of the 1970s, the percentage of the urban population rose again. The abolition of people's communes and the enforcement of market prices of agricultural products stimulated the major flow of the surplus population from rural to urban areas. The government had to permit rural migrants to register in cities of destination in 1984. Although a temporary identity card has been issued for migrant workers, rural hukou holders are denied administrative services such as social security, medical care, or education.

While Shanghai has been the largest city in China since 1950, Beijing was in fourth place in 1950 and 1955 (UNPD 2018). In Table 6.9, the ratio of Shanghai to Beijing is shown only after 1960, which expresses the first/second ratio. The ratio decreased in 1960–80, implying that Shanghai was more affected than Beijing by the Cultural Revolution. The ratio increased in 1980–2000, reflecting the impact of economic reform, but it fluctuated after 2000 owing to the similar population growth rates in two cities. Unlike other capital cities in Eastern Asia, Beijing did not experience a population decrease although the growth rate in 2015–20 dropped to almost zero.

6.8 Cross-Border Migration in Contemporary Eastern Asia

6.8.1 Overseas and Foreign Populations

The conclusion of WWII caused a large amount of cross-border migration in the late 1940s. As described in the previous chapter, more than six million Japanese repatriated to Japan by 1950. Combined with the postwar baby boom in 1947–49, the anxiety for overpopulation was serious. Fortunately, Japan could avoid the problem owing to the economic boom and fertility decline of the 1950s. There were more than two million Koreans in Japan in August 1945. Although the exact number is unknown, it is estimated that more than one million returned to the southern part of the Korean peninsula by 1950. It is also supposed that another one million migrated from Manchuria, China, and the northern Korea (Kim 1965; Kwon et al. 1975). In Taiwan, while 480,000 Japanese retreated, one million Mainlanders landed in 1949 when the KMT was defeated by the People's Liberation Army. In addition, there must have been a large amount of emigration from mainland China to Hong Kong, Macau, Southeastern Asia, and other parts of the world during the civil war and immediately after the foundation of the People's Republic of China.

The long-term trend of the overseas population is available only for Japanese and Koreans. As shown in Fig. 6.13, the population of overseas Japanese was 267,000 in 1970, which was only 4% of the 6.3 million in 1945. The number gradually increased to 1.41 million in 2019 but decreased to 1.31 million in 2022 due to the COVID-19 pandemic. The number of overseas Koreans suddenly increased from 2.32 million in 1990 to 4.83 million in 1991 because the population of Korean Chinese was included from this year. While the Japanese data count those who hold the Japanese nationality, Korean data include Korean descendants with foreign nationality. However, according to the Overseas Koreans Agency (2021), 2.51 million overseas Koreans in 2021 had the Korean nationality, which was almost twice as large as the number of overseas Japanese.

Figure 6.14 shows the change in percentage of the foreign population in Japan, Korea, and Taiwan. Refugees from DPR Korea are not included in "foreigners" in the Republic of Korea. Those from China, Hong Kong, and Macau were included only after 2012 in "foreigners" in Taiwan. The percentages of foreigners in Japan, Korea, and Taiwan are much lower than in developed countries in Europe and Northern America. Still, the proportion has increased since the late 1980s in Japan and since the late 1990s in Korea and Taiwan. The growth rate has been significantly higher in Korea and Taiwan than in Japan especially in the recent period. This may reflect the difference in economic conditions and governmental attitudes. Japan has been more reluctant in accepting foreign laborers than Korea or Taiwan. It was in the 2018 Immigration Control Act that the Japanese government declared to accept manual laborers. As a result, the foreign population counted in censuses considerably increased in 2015–20.

According to Table 6.12, the overseas Chinese have founded the largest community in Japan, followed by the Vietnamese and Koreans. It was after the Chinese

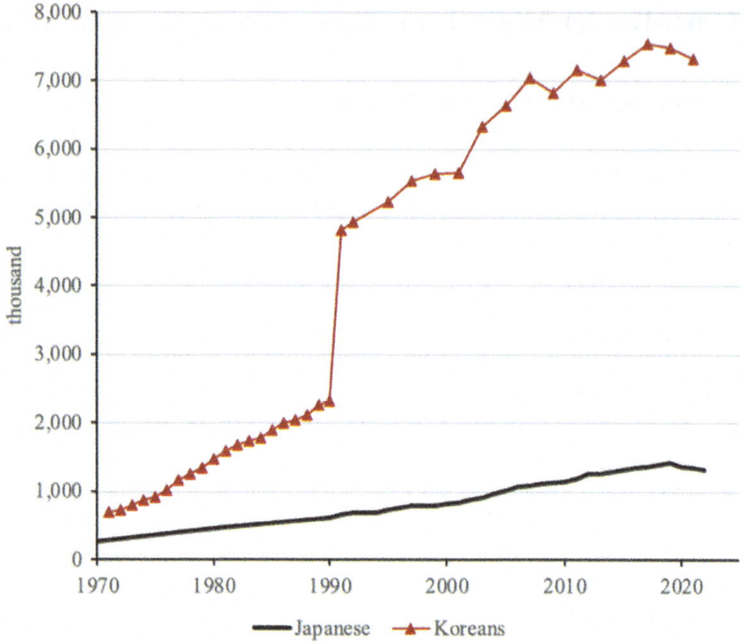

Fig. 6.13 Overseas Japanese and Koreans: 1970–2022. Ministry of Foreign Affairs of Japan (2023), Ministry of Foreign Affairs of Korea (2017), Overseas Koreans Agency (2021)

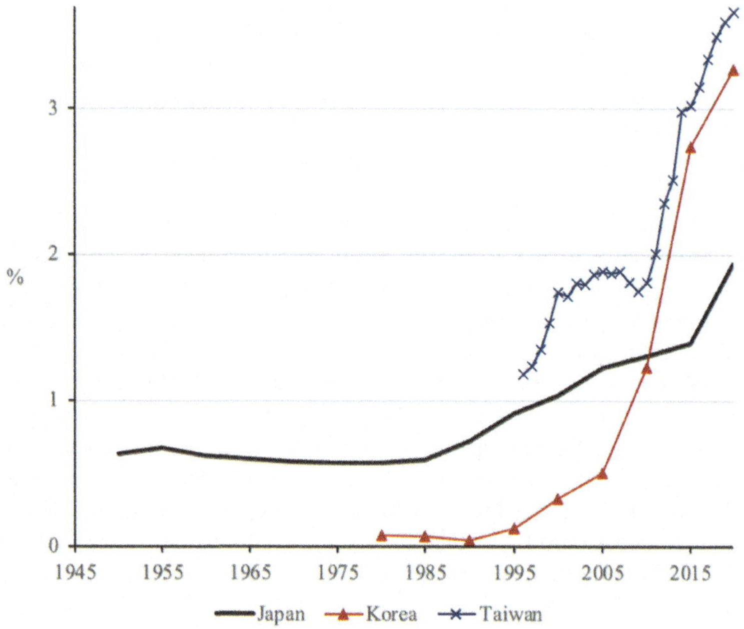

Fig. 6.14 Percentage of foreign population. Censuses, Ministry of the Interior

government permitted to travel abroad in 1986 that the entry of Chinese migrants to Japan radically increased. Japan has also attracted migrants from Southeastern Asia through historical relationships, governmental aids, private investments, overseas studentships, the industrial trainee program, the Memorandum of Understanding (MOU) on care workers, and cross-border marriages. Immigration from Brazil and Peru increased owing to the 1990 Immigration Control Act that allowed working visas for Japanese descendants in these countries. The Act also launched the industrial trainee program, while avoiding the Employment Permission System (EPS) as operated in Korea and Taiwan. Japan started accepting nurses and care workers from ASEAN countries based on the Economic Partnership Agreement (EPA) in 2004. The governmental effort to increase the number of overseas students officially aims at an improvement in international relationships. However, more direct purposes are to support local colleges and to employ part-time manual workers. In these "side-door" strategies, Japan avoided accepting low-skilled workers officially through an EPS program (Koido and Kamibayashi 2018). However, the 2018 Immigration Act opened door for foreign manual laborers by founding a new visa type for industries with serious labor shortage such as agriculture, fishery, restaurant, nursing care, cleaning, manufacturing, construction, etc. As a result, the registered foreign population increased from 2.73 million in 2018 to 2.93 million in 2019. Although the number decreased during the pandemic, it rose again and exceeded three million for the first time in 2022.

Table 6.12 Registered foreign population in Japan: 2020

	Persons	(%)
Total	3,075,213	(100.0)
China	761,563	(24.8)
Vietnam	489,312	(15.9)
Republic of Korea	411,312	(13.4)
Philippines	298,740	(9.7)
Brazil	209,430	(6.8)
Nepal	139,393	(4.5)
Indonesia	98,865	(3.2)
United States	60,804	(2.0)
Taiwan	57,294	(1.9)
Thailand	56,701	(1.8)
Myanmar	56,239	(1.8)
Peru	48,914	(1.6)
India	43,886	(1.4)
Sri Lanka	37,251	(1.2)
DPR Korea	25,358	(0.8)
Other	280,151	(9.1)

Ministry of Justice of Japan (2020)

Table 6.13 shows the foreign population in the 2020 census of Korea. Because Korea turned to a registration census after 2015, the number is equivalent with the registered foreign population. Both the largest and the second largest groups were from China. If Korean descendants were combined with non-Korean Chinese, immigrants from China would account for more than a half of the foreign population. Labor migration from Asian countries increased as of the late 1980s when the labor shortage in 3D (Dirty, Difficult, Dangerous) industries became prominent. As in Japan, Korea used to accept foreign manual laborers through the industrial trainee program started in 1993. However, Korea launched the EPS in 2004 and abolished the trainee program in 2006. In the EPS program, private agents are excluded, and recruitment is conducted by governmental agencies both in the sending country and in Korea. To achieve this, Korea has concluded the MOU with 16 countries: Thailand, Philippines, Sri Lanka, Vietnam, Indonesia, Mongolia, Pakistan, Uzbekistan, Cambodia, China, Bangladesh, Nepal, Myanmar, Kazakhstan, East Timor, and Laos (Imaizumi 2014). Eleven of them can be found in Table 6.13. The Korean Chinese are out of the frame of EPS and can more easily stay and work in Korea. In the permission system specified to Korean descendants in China and other countries, private agents work to recruit employees and conclude contracts with employers.

Table 6.14 shows the foreign population by nationality in Taiwan in 2020. The largest group is Indonesians, followed by Vietnamese, Filipinos, Thai, Chinese, and Malaysians. The EPS of Taiwan was founded in 1992 and has accepted manual laborers mainly from ASEAN countries. Although Mongolian workers are also acceptable, the number has been negligible since 2000. Because Taiwan does not

Table 6.13 Registered foreign population in Korea: 2020

	Persons	(%)
Total	1,695,643	(100.0)
Korean Chinese	541,337	(31.9)
China	207,764	(12.3)
Vietnam	199,950	(11.8)
Thailand	165,511	(9.8)
United States	59,646	(3.5)
Uzbekistan	58,000	(3.4)
Philippines	43,455	(2.6)
Cambodia	40,718	(2.4)
Nepal	39,385	(2.3)
Mongolia	37,064	(2.2)
Indonesia	34,675	(2.0)
Kazakhstan	27,035	(1.6)
Myanmar	25,609	(1.5)
Other	215,494	(12.7)

Statistics Korea

Table 6.14 Registered foreign population in Taiwan: 2020

	Persons	(%)
Total	797,122	(100.0)
Indonesia	253,570	(31.8)
Vietnam	239,039	(30.0)
Philippines	154,228	(19.3)
Thailand	64,017	(8.0)
China	60,451	(7.6)
Malaysia	22,726	(2.9)
Japan	15,914	(2.0)
United States	11,861	(1.5)
Republic of Korea	5,029	(0.6)
India	4,497	(0.6)
United Kingdom	2,657	(0.3)
Canada	2,568	(0.3)
Other	21,016	(2.6)

Ministry of the Interior

accept low-skilled workers from mainland China, the population of Chinese nationality is smaller than those from four ASEAN countries. This is contrasted with the amount and advantage of Korean Chinese in Korea.

There are no official statistics on the population of overseas Chinese with the nationality of People's Republic of China. The number including Chinese descendants is estimated to be 40 or 50 million (Chen 2013). Although the 2010 census of China reported the foreign population, the number seems to be too small compared with the overseas populations reported in the countries of origin. While the census counted 66,159 Japanese, the Ministry of Foreign Affairs of Japan (2011) reported that there were 131,534 Japanese on 1 October 2010 in China. The census count of 120,750 Korean nationality holders also seems too small if compared with 349,404 reported in the Ministry of Foreign Affairs of Korea (2017), although there is a seven-year interval. Anyway, the total number of 1,370,537 foreigners in the census report which accounted for only 0.1% of the total population must have been an underestimation.

In summary, although the foreign population in Japan started increasing earlier, Korea and Taiwan have already overtaken Japan. It is uncertain if the recent policy change allows Japan to exceed Korea and Taiwan again in the future. China is certainly a more closed to foreigners than Japan. As for overseas population, the ratio of overseas Koreans including foreign nationality holders to the domestic population in 2017 was 7,430,659/51,361,991 = 14.5%. If the denominator is replaced by the sum of the Republic and DPR Korea estimated by the Statistics Korea, the ratio will be 7,430,659/76,376,127 = 9.7%. This is apparently larger than the ratio of overseas Chinese that is supposed to be around 3%. Unlike in the modern period, the overseas

Japanese population today is only 1% of the homeland population. Taiwan reports only the number of overseas workers with the nationality of the Republic of China. According to the DGBAS (2021), the number decreased from 739,000 in 2019 to 501,000 in 2020 due to the corona pandemic. Because 500,000 oversea workers accounted for 2.2% of the domestic population, the ratio of oversea Taiwanese to the homeland population can be larger than that of China.

6.8.2 Cross-Border Marriages

As the foreign population increased in Japan, Korea, and Taiwan, cross-border marriage also increased. With economic development and female labor force participation, low-income men began to have difficulty marrying. The marriage squeeze was especially serious in rural areas. Marriage brokers appeared to satisfy the needs of men in developed countries by supplying single women from developing countries. Since immigration control in Japan, Korea, and Taiwan is still strict, marriage is the only way to obtain permanent residence qualification for unskilled women in developing countries (Tseng 2010).

Figure 6.15 shows the trend in percentage of cross-border marriage in Japan, Korea, and Taiwan. The nationwide data for China could not be found. The impact of the COVID-19 pandemic was less apparent in Japan. Taiwan recorded 31.9% at its peak in 2003. The percentage declined to 13.0% in 2015, although it rose again slightly until the pandemic. Of 21,216 cross-border marriages in 2019, 15,268 (72.0%) were between non-Taiwanese wives and Taiwanese husbands. There were 6580 wives from mainland China including Hong Kong and Macau, which accounted for 43.1% of all non-Taiwanese wives. The second largest group consisted of 5676 Vietnamese wives (37.2%), followed by 775 Indonesian wives (5.1%).

Cross-border marriages in Korea peaked at 42,356 (13.5%) in 2005, then decreased until 2015. Of 23,643 cross-border marriages in Korea in 2019, 17,687 (74.8%) were marriages with foreign wives. The largest group consisted of 6712 Vietnamese wives (37.9%), followed by 3649 Chinese wives including Korean Chinese (20.6%). Of 5956 foreign husbands, 1468 American men (24.6%) and 1407 Chinese men (23.6%) were dominant majorities.

The percentage of cross-border marriages in Japan did not rise as sharply as in Taiwan or Korea. The peak was in 2006 with 6.5% of all marriages. Of 21,919 cross-border marriages in 2019, 14,911 (68.0%) were between foreign wives and Japanese husbands. Chinese wives (4723 or 31.7%) were the largest group, followed by Filipino wives (3666 or 24.6%) and Korean wives (1678 or 11.3%). The largest group of foreign husbands consisted of 1764 Korean men (25.2%).

There are more marriages between wives from poor countries and husbands from rich countries than the opposite pairing. Thus, the wife/husband ratio in Table 6.15 is supposed to reflect economic superiority and inferiority. As far as the cross-border marriage is concerned, the Republic of Korea can be seen as a more developed country than Japan or Taiwan. In 2019, there were more cross-border marriages

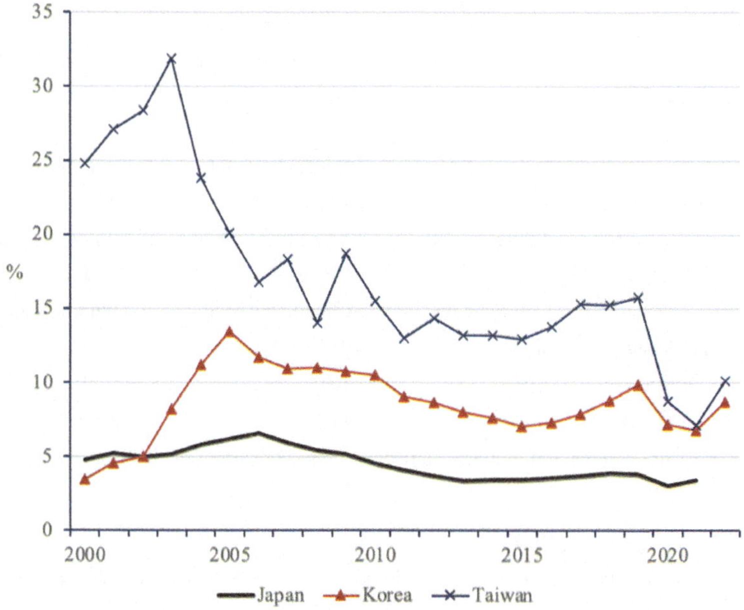

Fig. 6.15 Percentage of cross-border marriage. Ministry of Health, Labour and Welfare, Statsitics Korea, Ministry of the Interior

between Korean men and Japanese women than the opposite paring both in Japan and Korea. The contrast between Korea and Taiwan was more apparent. Although statistics in China were not available, cross-border marriages in 2019 suggested that the developmental level of China is still lower than that of Japan, Korea, and Taiwan. The low ratio in Korea suggests that many Chinese men including Korean Chinese achieved economic success in Korea.

6.9 Population of DPR Korea

Table 6.16 shows the population of DPR Korea since 1944. It is estimated that there were around 10 million residents in the domain of today's DPR Korea in 1944 (Kim et al. 2011). However, the official population publicized by the Central Bureau of Statistics in 1946 was about 9.3 million. This decrease was mainly caused by the flow into the south. The population of the north decreased further due to the Korean War. The death toll in the north exceeded one million, which was much larger than in the south where the population did not decrease (Kim 1965; Ishi 1972).

DPR Korea sustained very high population growth rates after the war throughout the 1960s. However, fertility started declining in the 1970s, despite the pro-natal attitude of the government. Moon (2011) attributed the decline to the frequent relocation

Table 6.15 Cross-border marriage couple by registered country: 2019

	Japanese wife	Japanese husband	Wife/husband
Registered in Korea	903	265	3.41
Registered in Taiwan	378	648	0.58
	Korean wife	Korean husband	Wife/husband
Registered in Japan	1,678	1,764	0.95
	Taiwanese wife	Taiwanese husband	Wife/husband
Registered in Korea	286	117	2.44
	Chinese wife	Chinese husband	Wife/husband
Registered in Japan	4,723	917	5.15
Registered in Korea	3,649	1,407	2.59
Registered in Taiwan	6,580	1,749	3.76

Ministry of Health, Labour and Welfare, Statistics Korea, Ministry of the Interior

Table 6.16 Population of DPR Korea

Year	Population (thousand)	Period	Annual growth rate (%)
1944	10,039[a]	1944–46	− 3.97
1946	9,257[b]	1946–49	1.30
1949	9,622[b]	1949–53	− 3.08
1953	8,491[b]	1953–60	3.48
1960	10,789[b]	1960–65	2.84
1965	12,408[b]	1965–70	3.33
1970	14,619[b]	1970–75	1.80
1975	15,986[b]	1975–80	1.59
1980	17,298[b]	1980–85	1.67
1985	18,792[b]	1985–93	1.53
1993	21,213[c]	1993–2008	0.84
2008	24,052[c]		

[a] Kim et al. (2011)
[b] Central Bureau of Statistics cited in Kim et al. (2011)
[c] Census

of male laborers and the mobilization of the female labor force. After the collapse of the former Soviet Union, DPR Korea suffered from economic difficulty. The flood in 1995 seriously damaged agricultural production, and a large number of people died from the long-term famine in 1996–2000. As mentioned above, estimates of excess deaths ranged between 270,000 and 3,500,000.

DPR Korea conducted censuses in 1993 and 2008 with aid from the UNFPA. While the UNPD (2022) assumed that these census populations were underestimated, the Statistics Korea gave estimates slightly smaller than the census results. The estimated population for 2020 by the UNPD and the Statistics Korea were 25.9 million and 25.4 million, respectively. While the result of UNPD suggested that the population of DPR Korea will start declining in 2034, the Statistics Korea showed that the population decline will start in 2030.

The TFR values based on the censuses were 2.13 in 1993 and 2.00 in 2008 (Kim et al. 2011). It is supposed, then, that fertility was at the below-replacement level during the famine and did not return to the replacement level. The life expectancy at birth based on the 2008 census was 64.8 for males and 71.9 for females. These were close to such developing countries as Indonesia, the Philippines, or Bangladesh.

Residential mobility is still strictly limited in DPR Korea. In addition, there is a political differentiation between Pyongyang residents and other people. Since the northern economy grew faster than the southern economy in the 1960s, urbanization was also faster in the north. However, the planned economy that did not stimulate innovation reached its limit in the 1970s. Thus, the percentage of the urban population was 60.6% in 2008, which is not much an increase from 59.1% in 1985 (Kim et al. 2011). It is known that DPR Korea sends several tens of thousands of laborers abroad to obtain foreign currencies. The cumulative number of refugees to the south exceeded 30,000 in 2016. Then, the total number of overseas North Koreans may not exceed 100,000.

6.10 Summary and Conclusion

After the end of WWII, migration on a major scale occurred, in addition to further warfare in Eastern Asia. More than six million repatriates returned to Japan, deteriorating the already disordered economic system. The divided occupation of Korea resulted in the war in 1950, and this division was perpetuated by the Cold War. The civil war in China ended with the victory of the communist army and one million Mainlanders appeared as a new ethnic group in Taiwan. Even after the foundation of the People's Republic of China, political malfunctions such as the Great Leap Forward and the Cultural Revolution caused further excess mortality and socioeconomic confusions.

While Japan had been the front-runner of Westernization and modernization in Asia since the nineteenth century, it was also Japan that began rapid economic growth in the earliest period after WWII. Taiwan followed Japan with its competitive agricultural products and light industries. Although the takeoff of Korea was slow because of the war, Korea eventually surpassed Taiwan with the advantage of a larger scale of economy including heavy industries. The development of China was delayed by political failures and the shortcomings of the planned economy.

The explanatory power of the economic development continued throughout the twentieth century. Japan led most demographic changes and China was behind Korea and Taiwan. While Japan reached the replacement level of fertility in the late 1950s, Korea and Taiwan achieved the fertility transition in the mid-1980s and China in the early 1990s. The life expectancy at birth was 76.1 years in Japan in 1980, which was the longest worldwide. Korea and Taiwan arrived at this level in the early 2000s, and China in early 2010s. The total dependency ratio started declining in the early 1950s in Japan, in the early 1960s in Taiwan, in the late 1960s in Korea, and in the early 1970s in China. The dependency ratio began to rise due to the population aging in the late 1990s in Japan and in 2010–15 in Korea, Taiwan, and China. The age at marriage and the divorce rate were higher in Japan than in Korea and Taiwan until the end of the twentieth century. Japan was the leader of household changes still in 2020 in terms of the average household size and the propensity to live alone. According to the UNPD (2018), the percentage of the urban population in 2015 was 91.4% in Japan, 81.6% in Korea, 76.9% in Taiwan, and 55.5% in China. Although the data for Taiwan before 1996 were not found, it is inferred that the percentage of foreign population was higher in Japan than in Korea or Taiwan until the early 1990s.

After the turn of the century, the developmental level lost its explanatory power on some demographic aspects. The age at marriage and the divorce rate are higher in Korea and Taiwan than in Japan today. The TFRs of Korea and Taiwan have been lower than that of Japan since the early 2000s. As a result, Korea and Taiwan will override Japan in terms of population aging by the 2050s. The tempo of urbanization and the magnitude of concentration have been so drastic in Korea that the partial relocation of capital functions was conducted in 2012. The percentage of foreign population in Taiwan was already higher than Japan in the mid 1990s and that of Korea overtook Japan in the 2015 census. Japan's demographic advancement will be further eroded because the living standard of Korea and Taiwan have caught up Japan. As of 1 October 2023, the home page of the OECD. Stat showed that the GDP per capita (US dollar base) of Korea exceeded Japan in 2018, although the IMF World Economic Outlook Database estimated that the GDP per capita of Japan was still slightly higher than Korea or Taiwan in 2020.

China is the world's second largest economy today. According to the IMF, the GDP of China was about three times as large as Japan in 2020. This means that the GDP per capita of China was less than three tenth of Japan. China failed again to synchronize demography and economics in the second half of the twentieth century. While fertility decline in China was delayed a decade or so compared to the Republic of Korea and Taiwan, economic development started significantly later than in those capitalist societies. Unlike Korea and Taiwan that made use of the declining dependency ratio since the 1960s, China was not ready to take the demographic window of opportunity in the 1970s. The slow start of economic development has resulted in the problem of early aging. Considering the recent economic recession, it seems difficult for China to catch up more advanced economies in Eastern Asia.

References

Ato, Makoto. 1997. Nihon-no choshosanka gensho-to kachikan hendo kasetsu. *Journal of Population Problems* 53 (1): 3–20 (in Japanese).

Chang, Se-Hun. 2002. Doshihoa. In *Hanguk-eui Ingu*, eds. Kim Doo-Sub, Park Sang-Tae, and Eun Ki-Soo, 495–523. National Statistics Office (in Korean).

Chen, Laixing. 2013. Imin-to kakyo. In *Chugoku Keizaishi*, ed. Okamoto Takashi, 232–234. Nagoya University Press (in Japanese).

Choi, Eun-Young, Park Se-Kyung, Lee Sam-Sik, Cho Nam-Hoon and Choi Byung-Ho. 2005. *Hanguk-eui Jeochulsan-gwanryeon Sahoegyeongjejeok Yoin-gwa Jeongchaek-yeogeon*. Korea Institute for Health and Social Affairs (in Korean).

Dikoetter, Frank. 2010. *Mao's Great Famine: The History of China's Most Devastating Catastrophe, 1958–1962*. Walker & Co.

Directorate—General of Budget, Accounting and Statistics, Executive Yuan. 2021. *107-nian Guoren Fuwaiguo Gongzuo Renshu Tongji Jieguo* (in Chinese).

Eckert, Carter J. 1991. *Offspring of Empire, The Koch'ang Kims and the Colonial Origins of Korean Capitalism 1876–1945*. University of Washington Press.

Gender Equality Committee of the Executive Yuan. 2021. *2021-nian Woguo Xingbie Bupingdeng Zhishu GII Fabu Shuoming*. https://gec.ey.gov.tw/File/60A944416FD7AB63?A=C (seen on 4 Aug 2023) (in Chinese).

Gender Equality Committee of the Executive Yuan. 2022. *2022-nian Woguo Xingbie Luocha Zhishu GGI Fabu Shuoming*. https://gec.ey.gov.tw/File/43946B90633D76A5?A=C (seen on 4 Aug 2023) (in Chinese).

Guo, Zhigang, and Gu Baochang. 2014. China's low fertility: evidence from the 2010 census. In *Analysing China's Population—Social Change in a New Demographic Era*, eds. Isabelle Attane and Gu Baochang, 15–35. Springer.

Guo, Zhigang, Stuart Gietel-Basten and Baochang Gu. 2018. The lowest fertility rates in the world? Evidence from the 2015 Chinese 1% sample census. *China Population and Development Studies* 2 (3): 245–258.

Horiuchi, Shiro. 2010. Nihonjin-no jumyo encho: Yoin-to tenbo. *Journal of Population Problems* 66 (3): 40–49 (in Japanese).

Imaizumi, Shinya. 2014. Higashi Asia-ni okeru gaikokujin koyo hosei-no kosatsu. In *Higashi Asia-ni okeru Imin Rodosha-no Hoseido – Okuridashikoku-to Ukeirekoku-no Kyotsu Kiban-no Kochiku-ni Mukete*, ed. Yamada Miwa. Institute of Developing Economies (in Japanese).

Inglehart, Ronald, and Christian Welzel. 2008. *Modernization, Cultural Change, and Democracy—The Human Development Sequence*. Cambridge University Press.

Ishi, Yoshikuni. 1972. *Kankoku-no Jinko Zoka-no Bunseki*. Keiso Shobo (in Japanese).

Ishi, Hiroyuki. 2018. *Kansensho-no Sekaishi*. Kadokawa (in Japanese).

Kelley, Allen C. 1988. Economic consequences of population change in the third world. *Journal of Economic Literature* 26 (4): 1685–1728.

Kim, Doo-Sub, Choi Min-Za, Jun Kwang-Hee, Lee Sam-Sik, and Kim Hyung-Seok. 2011. *Bukhan Ingu-wa Ingu Census*. National Statistics Office (in Korean).

Kim, Cheol. 1965. *Kankoku-no Jinko–to Keizai*. Iwanami Shoten (in Japanese).

Koido, Akihiro, and Chieko Kamibayashi. 2018. "Nihon shakai-to kokusai imin – ukeire ronso 30-nengo-no genjitsu"-ni yosete. *Japanese Sociological Renew* 68 (4): 468–478 (in Japanese).

Koike, Shiro. 2017. Tokyoto kubu-ni okeru 'toshinkaiki'-no jinkogakuteki bunseki. *The Journal of Population Studies* 53: 23–45 (in Japanese).

Kwon, Tai-Hwan, and Kim Doo-Sub. 2002. *Ingu-eui Ihae*. Seoul National University Press (in Korean).

Kwon, Tai-Hwan, Lee Hae-Young, Chang Yunshik, and Yu Eui-Young. 1975. *The Population of Korea*. Population and Development Studies Center, Seoul National University.

Lee, Meilin, and Lin Yu-Hsuan. 2016. Transition from anti-natalist to pro-natalist policies in Taiwan. In *Low Fertility, Institutions, and their Policies: Variations across Industrialized Countries*, eds. Ronald R. Rindfuss and Minja Kim Choe, 259–281. Springer.

Lee, Sang-Hyop, Andrew Mason, and Donghyun Park. 2012. Overview: Why does population aging matter so much for Asia? Population aging, economic growth, and economic security in Asia. In

Aging, Economic Growth, and Old-Age Security in Asia, eds. Park Donghyun, Lee Sang-Hyop, and Andrew Mason, 1–31. Edward Elgar.
Lee, Ronald D. 2007. *Global Population Aging and its Economic Consequences.* AEI Press.
Lesthaeghe, Ron. 2010. The Unfolding story of the second demographic transition. *Population and Development Review* 36 (2): 211–251.
Mason, Andrew. 2001. Population and economic growth in East Asia. In *Population Changes and Economic Development in East Asia*, ed. Andrew Mason, 1–30. Stanford University Press.
McDonald, Peter. 2000. Gender equity in theories of fertility transition. *Population and Development Review* 26 (3): 427–440.
Messer, Ellen, and Laurie F. DeRose. 1998. Food shortage. In *Who's Hungry? And How Do We Know? Food Shortage, Poverty, and Deprivation*, eds. Laurie DeRose, Ellen Messer, and Sara Millman, 53–91. United Nations University Press.
Ministry of Foreign Affairs of Japan. 2011. *Kaigai Zairyu Hojin Ninzu Chosa Tokei 2011* (in Japanese).
Ministry of Foreign Affairs of Japan. 2023. *Kaigai Zairyu Hojin Chosa Tokei 2023* (in Japanese).
Ministry of Foreign Affairs of Korea. 2017. *Jaeoe Dongpo Hyeonhyang 2017* (in Korean).
Ministry of Justice of Japan. 2020. *Zairyu Gaikokujin Tokei 2020* (in Japanese).
Ministry of the Interior. 2009. *Renkou Zhengcc Ziliao Huiji* (in Chinese).
Moon, Ho-Il. 2011. *Chosen Minshushugi Jinmin Kyowakoku-no Jinko Hendo – Jinkogaku-kara Yomitoku Chosen Shakaishugi.* Akashi Shoten (in Japanese).
National Development Council, Republic of China. 2022. *Zhonghua Mínguo Renkou Tuigu* (in Chinese).
National Health Commission of the People's Republic of China. 2020. *2019-nian Woguo Weisheng Jiankang Shiye Tongji Gongbao* (in Chinese).
National Health Commission of the People's Republic of China. 2021. *2020-nian Woguo Weisheng Jiankang Shiye Tongji Gongbao* (in Chinese).
National Health Commission of the People's Republic of China. 2022. *2021-nian Woguo Weisheng Jiankang Shiye Tongji Gongbao* (in Chinese).
National Institute of Population and Social Security Research (IPSS). 2023a. *Jinko Tokei Shiryoshu 2023 Kaiteiban* (in Japanese).
National Institute of Population and Social Security Research (IPSS). 2023b. *Nihon-no Shorai Suikei Jinko: 2021–2070* (in Japanese).
Nippo, Jinmin. 2015. Chugoku-no shinseiji danjo hiritsu-ga 6-nen renzoku-de kaizen-mo imada unbalance tsuzuku, 9 Feb 2015. http://j.people.com.cn/n3/2016/1012/c94475-9125737.html (seen on 6 Aug 2023) (in Japanese).
Office of the Leading Group of the State Council for the Seventh National Population Census. 2021. *2020-nian Dai7-ci Quanguo Renkou Puzha Shuyo Shuju.* China Statistics Press (in Chinese).
Omran, Abdel R. 1971. The epidemiological transition. *Milbank Memorial Fund Quarterly* 49 (4): 509–508.
Onishi, Hiroshi. 2014. *Senshinkoku Kankoku-no Yuutsu – Shoshi-koreika, Keizaikakusa, Global-ka.* Chuko Shinsho (in Japanese).
Overseas Koreans Agency. 2021. *Jaeoe Dongpo Hyeonhyang 2021* (in Korean).
Ping, Wen-Meng. 2009. *Chugoku-no Jinko Ido-to Shakaiteki Genjitsu.* Toshindo (in Japanese).
Rindfuss, Ronald R., and Minja Kim Choe. 2015. Diversity across low-fertility countries: an overview. In *Low and Lower Fertility: Variations across Developed Countries*, eds. Ronald R. Rindfuss, and Minja Kim Choe, 1–13. Springer.
Smith, P.C. 1980. Asian marriage patterns in transition. *Journal of Family History* 5 (1): 58–96.
Statistics Korea. 2021. *Jangrae Ingu Chugye: 2020–2070-nyeon* (in Korean).
Statistics Korea. 2022. *Jangrae Ingu Chugye Shidopyeon: 2020–2050-nyeon* (in Korean).
Suzuki, Toru. 2021. Population aging, labor force participation, and family structure in the Republic of Korea. In *Employment, Retirement and Lifestyle in Ageing East Asia*, ed. Ma XinXin, 241–256. Palgrave Macmillan.
Suzuki, Toru. 2013. *Low Fertility and Population Aging in Japan and Eastern Asia.* Springer.

Tian, Xueyuan. 2007. *21-shiji Zhongguo Renkou Fazhan Zhanlue Yanjiu*. Social Science Academic Press (in Chinese).
Tseng, Yen-Fen. 2010. Marriage migration to East Asia—Current issues and propositions in making comparisons. In *Asian Cross-border Marriage Migration—Demographic Patterns and Social Issues*, eds. Yang Wen-Shan, and Lu Melody Chia-Wen, 30–45. Amsterdam University Press.
United Nations Population Division. 2022. *World Population Prospects, The 2022 Revision*.
United Nations Population Division. 2018. *World Urbanization Prospects, The 2018 Revision*.
van de Kaa, Dirk. 1987. Europe's second demographic transition. *Population Bulletin* 42 (1): 1–59.
Vogel, Ezra F. 1991. *The Four Little Dragons*. Harvard University Press.
Wakabayashi, Keiko. 1989. *Chugoku-no Jinko Mondai*. Tokyo University Press (in Japanese).
Wakabayashi, Keiko, and Nie, Haisong. 2012. *Chugoku Jinkou Mondai-no Nenpu-to Tokei: 1949–2012-nen*. Ochanomizu Shobo (in Japanese).
Wakabayashi, Keiko. 1996. *Gendai Chugoku-no Jinko Mondai-to Shakai Undo*. Shinyosha (in Japanese).

Chapter 7
Concluding Remarks

This book showed that some important population issues today have long historical roots. The emergence of the lowest level of fertility in Korea and Taiwan was attributed to the family pattern that was established during the pre-modern era. The fertility decline in Japan was within the range of Europe because Japan was already similar to Europe in the preindustrial period. On the other hand, Confucian families in Korea and Taiwan substantially differed from feudal families in Japan or Europe in terms of gender and parent–child relations. The drastic change in political, educational, and occupational systems in the postindustrial period caused a wider gap between the family and non-family systems than in other advanced society, which resulted in the extremely low level of fertility. The TFRs reported in the 2000 and 2010 census of China were supposed to be underestimated. On the other hand, the beginning of population decline in recent China suggests that fertility after 2010 has been as low as Korea and Taiwan while mortality has been significantly higher.

China's population growth began in the late seventeenth century, which was earlier than in England. Population growth continued until the late eighteenth century, supported by the increase in cultivated lands. However, China could not be the pioneer of the industrial revolution, and its population collapsed in the nineteenth century. The failure to synchronize population growth and economic development appeared again in the second half of the twentieth century. In addition to the drawback of the planned economy, political disasters such as the Great Leap Forward and the Cultural Revolution blocked the economic development. Thus, China could not start the economic development when the demographic window of opportunity opened in the 1970s. The problem of early aging derived from such a delayed development.

Although Korea and Taiwan were colonized by Japan, the modes of development were completely different. While the rice monoculture in Korea could not absorb the rural surplus population, Taiwanese villagers could stay home producing sugar, tobacco, tea, camphor, and various fruits, in addition to rice. As a result, more than 95% of Taiwanese lived in their homeland in 1930. While there were more than two million Koreans in Japan in 1945, the number of Taiwanese was only around 35,000.

T. Suzuki, *Comparative Population History of Eastern Asia*,
https://doi.org/10.1007/978-981-99-9367-3_7

The different modes of development caused different magnitudes of urbanization in the latter half of the twentieth century. Because the urbanization was so compressed and the concentration in the capital area was so intense, the capital relocation was eventually executed in the 2010s. Such huge amount of diaspora from Korea and the compressed urbanization were stimulated by the rice monoculture economy under Japan's rule.

Japan has been the front-runner in Eastern Asia, followed by Korea and Taiwan, and China has been behind them. Although demographic indicators coincided with the developmental level for a long time, some indices started to deviate from the developmental order. The deviation from the developmental level implies that all social changes, not necessarily limited to demographic changes, are affected by various factors. Because economic development is the primary driving force of demographic transition, urbanization, and cross-border migration, these changes were strongly determined by the developmental level at the early stage. As time went by, however, various factors hidden behind economic development began to exhibit their effects. In this book, various deviations were interpreted from different determinants. The change in the order of fertility levels was interpreted as resulting from the cultural disorder of different family patterns. The different tempo of urbanization was interpreted as resulting from different modes of agriculture. The increase in foreign population was interpreted as resulting from different governmental attitudes.

As for the premodern history, it was shown that the damage of epidemic became less dangerous in recurrent epidemics by the co-evolution of disease orgasm and human hosts. Even for the plague, excess death rates in the pandemic that started in 1894 were 1% or less in each country. The deaths from acute infectious diseases dramatically declined in the modern period owing to the nutritional improvement, medical innovation, and public health programs. In Japan, the death toll of the Spanish flu in the 1920s was smaller than the normal number of deaths by tuberculosis. The tuberculosis was overcome after WWII by streptomycin and BCG vaccine. While a reoccurrence of highly toxic influenza pandemic was worried in the contemporary era, what happened was the COVID-19 pandemic by the less lethal but highly infectious corona virus. Although this pathogen had a serious damage on socioeconomic activities through the limitation of mobility, the death rate was much smaller than the pandemics in the past.

It was also shown that the demographic impact of famine decreased under the economic development and wealth accumulation. Population declines by great famines were around 10% in the seventeenth-century Korea and 5% in the eighteenth-century Japan. After the modernization, great famines in Eastern Asia were caused by malfunctions of governments. The great Chinese famine during the Great Leap Forward period occurred due to the unrealistic production plans and inappropriate labor mobilization by the government. The famine in the DPR Korea happened because the government could not respond adequately to the collapse of the Soviet Union and did not request an international aid promptly.

While the damage of war drastically expanded in the twentieth century, some wars in the preindustrial period resulted in catastrophic population disasters. For example, the Mongol invasion to Korea caused a disastrous demographic crisis because the war accompanied not only famine and epidemics but also the labor extrapolation for further expeditions. The foundation of Mongol empire triggered the plague pandemic in the whole Eurasian continent. The Red Turban Rebellions at the end of the Yuan dynasty also accompanied the plague epidemics and famines, destroying a large part of the Chinese population. Japan was free from an invasion throughout the premodern era. Although Eastern Asian countries including Japan suffered seriously from a series of wars in the 1940s and 1950s, damages were still smaller than European countries.

It was assumed in this book that the population density was 1.0/km^2 in Eastern Asian hunter-gatherer societies. In the case of Japan, the population density rose to 89/km^2 during 2800 years of agrarian period and hit the peak at 343/km^2 after 150 years of growth in industrial period. While the maximum population density that was achieved recently in the Korean peninsula was about the same as Japan, the peak was considerably higher in Taiwan and lower in China. The population decline started earlier in Southern and Eastern Europe than in Eastern Asia. The maximum populations experienced in these regions are assumed to be smaller than the carrying capacity, namely the maximum size that the land area can sustain. In Malthusian theory, it was assumed that when a population approaches to the ecological limit, the living standard drops to the subsistence level and the population growth is checked by rising mortality. Fortunately, the population declines in Southern/Eastern Europe and Eastern Asia were driven by declining fertility.

No developed country in Europe, Northern America, and Eastern Asia could sustain the replacement level of fertility since the 1990s. This implies that the lifestyle of developed countries in the postindustrial period is incapable of reproduction of population. Reasons may include the worsened labor market condition due to the slow economic growth, the rising cost of childcare and education, the abandonment of the merit of division for gender equity, the individualization rejecting the traditional and religious authorities, and the emphasis on autonomy and privacy denying the scale merit. A population with below-replacement fertility eventually starts to decline if not compensated by immigration. Some Eastern European countries experienced the population decline owing to the diaspora stimulated by the political and economic disorder in the 1990s. Most Southern European countries experienced lowest-low fertility in the 1990s and the population decline started around 2010. Eastern Asian countries exhausted the momentum of the past population growth during 30–35 years of below-replacement fertility. In Northern and Western Europe, some countries including Germany, Austria, Belgium, and Finland sustained the population with immigrants despite the rate of natural increase was negative in 2020. In other Northern/Western European and English-speaking advanced countries, the natural increase was still positive owing to relatively high fertility and the young age structure by accepting immigrants.

Developed countries that cannot sustain the population without immigration resemble to cities in the preindustrial period. Because the rate of natural increase was negative in premodern cities, the population was sustained by absorbing in-migrants from rural areas. This pattern of domestic migration was reproduced in the global pattern and developed countries today are absorbing immigrants from developing countries like a blackhole. Countries in Southern/Eastern Europe and Eastern Asia are suffering the population decline because of the insufficient suction power. The issue of migration has caused serious political conflicts and nourished the nationalism in Europe and Northern America. On the other hand, most countries in Southern/Eastern Europe and Eastern Asia cannot accept enough immigrants and are experiencing stagnation due to labor shortage. As can be seen in the recent policy change of Japan, the competition to attract foreign labor force among Eastern Asian advanced societies will be intensified hereafter. The problem of early aging will be worsened if Chinese workers are absorbed into these societies.

Character List

Propper nouns and other words appeared in texts, tables, and figures were chosen and written in Eastern Asian writing system. Zodiac names in Japanese were written in hiragana because it is difficult even for Japanese to read them if written in Chinese characters. Korean words were written in hangul because Chinese characters are rarely used today. Due to the space limitations, only traditional characters were shown for Chinese.

English	Category	Japanese	Korean	Chinese
Aichi	Prefecture	愛知	아이치	愛知
ak-byeong	Epidemic	悪病	악병	惡病
Akita	City, prefecture	秋田	아키타	秋田
An Lushan	Person	安禄山	안록산	安祿山
Anhui	Province	安徽	안후이	安徽
Ansei	Era	安政	안세이	安政
Anyang	City	安養	안양	安養
Aomori	City, prefecture	青森	아오모리	青森
Ashikaga Takauji	Person	足利尊氏	아시카가 다카우지	足利尊氏
Ashikaga Yoshimitsu	Person	足利義満	아시카가 요시미N	足利義滿
Asuka	Era	飛鳥	아스카	飛鳥
Azuchi-Momoyama	Era	安土桃山	아즈치 모모야마	安土桃山
bakufu	Organization	幕府	막부	幕府
Beakje	Kingdom	百濟	백제	百濟
Beijing	City	北京	베이징	北京
Beiping	City	北平	베이핑	北平
Benhan	People	弁韓	변한	弁韓
bing-zi	Year	ひのえね	병자	丙子

(continued)

T. Suzuki, *Comparative Population History of Eastern Asia*,
https://doi.org/10.1007/978-981-99-9367-3

(continued)

English	Category	Japanese	Korean	Chinese
Book of Han	Literature	漢書	한서	漢書
Book of the Later Han	Literature	後漢書	후한서	後漢書
Book of Wei	Literature	魏書	위서	魏書
Bucheon	City	富川	부천	富川
Busan	City	釜山	부산	釜山
Buyeo	People	扶余	부여	扶餘
byeong-ja	Year	ひのえね	병자	丙子
Cao Wei	State	曹魏	조위	曹魏
Cefu Yuangui	Literature	冊府元亀	책부원귀	冊府元龜
chaebol	Organization	財閥	재벌	財閥
Chagang	Province	慈江	자강	慈江
Chahar	Province	察哈爾	차하얼	察哈爾
Changhua	County	彰化	장화	彰化
Changjiang	River	長江	장강	長江
Changwon	City	昌原	창원	昌原
Chen Sheng	Person	陳勝	진승	陳勝
Chengshi	City	城市	성시	城市
Chiang Kai-shek	Person	蒋介石	장제스	蔣介石
Chiayi	City, county	嘉義	자이	嘉義
Chiba	City, prefecture	千葉	지바	千葉
Chikuma	River	千曲	치쿠마	千曲
Chinghai	Province	青海	칭하이	青海
Chongqing	City	重慶	충칭	重慶
Chongzhen	Emperor	崇禎帝	숭정제	崇禎帝
Chu	State	楚	초	楚
Chun Doo-Hwan	Person	全斗煥	전두환	全斗煥
Chungbuk	Province	忠北	충북	忠北
Chungcheong	Province	忠清	충청	忠清
Chungnam	Province	忠南	충남	忠南
da-ji	Famine	大飢	대기	大饑
dae-gi	Famine	大飢	대기	大饑
Daegu	City	大邱	대구	大邱
Daehan Minguk	Nation	大韓民国	대한민국	大韓民國
Daejeon	City	大田	대전	大田
Daifang	Commandery	帯方	대방	帶方
daimyo	Feudal lord	大名	다이묘	大名

(continued)

(continued)

English	Category	Japanese	Korean	Chinese
Dangun	Person	檀君	단군	檀君
Danseong	County	丹城	단성	丹城
Daoguang	Emperor	道光帝	도광제	道光帝
Daoyi	Village	道義屯	도의둔	道義屯
dayi	Epidemic	大疫	대역	大疫
Deng Xiaoping	Person	鄧小平	덩샤오핑	鄧小平
Di	People	氐	저	氐
Dong	City	洞	동	洞
Donghualu	Literature	東華録	동화녹	東華錄
Dungan Revolt	Rebellion	回民蜂起	둥간혁명	同治陝甘回乱
Eastern Jin	State	東周	동주	東周
Eastern Wu	State	東呉	동오	東吳
Edo	City	江戸	에도	江戶
Ehime	Prefecture	愛媛	에히메	愛媛
Emishi	People	蝦夷	에조	蝦夷
Eub	Town	邑	읍	邑
Eul-byeong Famine	Famine	乙丙大飢饉	을병대기근	乙丙大饑荒
eul-hae	Year	きのとい	을해	乙亥
Former Zhao	Kingdom	前趙	전조	漢趙
Fujian	Province	福建	푸젠	福建
Fujiwara	Person	藤原	후지와라	藤原
Fukui	City, prefecture	福井	후쿠이	福井
Fukuoka	City, prefecture	福岡	후쿠오카	福岡
Fukushima	City, prefecture	福島	후쿠시마	福島
Fushimi	Castle	伏見城	후시미성	伏見城
Gabo	Year	きのえうま	갑오	甲午
Ganghwa	Island	江華島	강화도	江華島
Gangwon	Province	江原	강원	江原
Gansu	Province	甘粛	간d	甘肅
Gaya	Kingdom	伽耶	가야	伽耶
geng-xu	Year	かのえいぬ	경술	庚戌
Genji	Clan	源氏	겐지	源氏
Genpei War	War	源平合戦	겐페이전쟁	源平合戰
Genroku	Era	元禄	겐로쿠	元祿
gi	Famine	飢	기	饑
Gifu	City, prefecture	岐阜	기후	岐阜
Gija	Person	箕子	기자	箕子

(continued)

(continued)

English	Category	Japanese	Korean	Chinese
Goguryeo	Kingdom	高句麗	고구려	高句麗
Gojong	King	高宗	고종	高宗
Gojoseon	Era	古朝鮮	고조선	古朝鮮
Goryeo	Dynasty	高麗	고려	高麗
Goryeosa	Literature	高麗史	고려사	高麗史
Guangdong	Province	広東	광둥	廣東
Guangxi	Province	広西	광시	廣西
Guangzhou	City	広州	광저우	廣州
Guizhou	Province	貴州	구이저우	貴州
Gun	Rural	郡	군	郡
Gunma	Prefecture	群馬	군마	群馬
Gwangju	City	光州	광주	光州
Gyeong-shin Famine	Famine	庚辛大飢饉	경신대기근	庚辛大饑荒
gyeong-sul	Year	かのえいぬ	경술	庚戌
Gyeongbuk	Province	慶北	경북	慶北
Gyeonggi	Province	京畿	경기	京畿
Gyeongju	City	慶州	경주	慶州
Gyeongnam	Province	慶南	경남	慶南
Gyeongsang	Province	慶尚	경상	慶尚
Hainan	Island, province	海南	하이난	海南
Hakka	Language	客家	하카	客家
Hambuk	Province	咸北	함북	鹹北
Hamgyong	Province	咸鏡	함경	鹹鏡
Hamnam	Province	咸南	함남	鹹南
Han	Dynasty	漢	한	漢
Han Chinese	People	漢族	한족	漢族
Han Koreans	People	韓民族	한민족	韓民族
Han River	River	漢江	한강	漢江
Hangzhou	City	杭州	항저우	杭州
Hankou	City	漢口	한커우	漢口
Harbin	City	ハルビン	하얼빈	哈爾濱
Hayato	People	隼人	하야토	隼人
Hebei	Province	河北	허베이	河北
Heian-kyo	City	平安京	헤이안쿄	平安京
Heijo-kyo	City	平城京	헤이조쿄	平城京
Heike	Clan	平家	헤이케	平家
Heilongjiang	Province	黒竜江	헤이룽장	黑龍江

(continued)

(continued)

English	Category	Japanese	Korean	Chinese
Heisei	Era	平成	헤이세이	平成
Henan	Province	河南	허난	河南
Heo Jun	Person	許浚	허준	許浚
Himiko	Queen	卑弥呼	히미코	卑彌呼
hinoe-ne	Year	ひのえね	병자	丙子
hinoeuma	Year	ひのえうま	병오	丙午
Hiroshima	City, prefecture	広島	히로시마	廣島
Hoei	Era	宝永	호에이	寶永
Hokkaido	Prefecture	北海道	홋카이도	北海道
Hoklo	People	福佬人	민난인	閩南人
Hokuriku	Region	北陸	호쿠리쿠	北陸
Hong Gyeong-Rae	Person	洪景来	홍경래	洪景來
Hong Kong	City	香港	홍콩	香港
Horeki	Era	宝暦	호레키	寶曆
Hsinchu	City, prefecture	新竹	신주	新竹
Huai	River	淮河	회하	淮河
Huai	Emperor	懐帝	회제	懷帝
Hualien	County	花蓮	화롄	花蓮
Hubei	Province	湖北	후베이	湖北
Huguang	Region	湖広	후광	湖廣
Hui	People	回族	회족	回族
hukou	Registration	戸口	호구	戶口
Hunan	Province	湖南	후난	湖南
Hwanghae	Province	黄海	황해	黃海
Hyogo	Prefecture	兵庫	효고	兵庫
Ibaraki	Prefecture	茨城	이바라기	茨城
ie	Family	イエ (家)	집	家
inamegasa	Disease	稲目瘡	이나메가사	稲目瘡
Incheon	City	仁川	인천	仁川
Iryeon	Person	一然	일연	一然
Ishikawa	Prefecture	石川	이시카와	石川
Iwate	Prefecture	岩手	이와테	岩手
jang-yeok	Epidemic	瘴疫	장역	瘴疫
Jeju	Island, province	済州	제주	濟州
Jeolla	Province	全羅	전라	全羅
Jeonbuk	Province	全北	전북	全北
Jeonnam	Province	全南	전남	全南

(continued)

(continued)

English	Category	Japanese	Korean	Chinese
Jeulmun	Culture	櫛文	즐문	櫛文
ji	Famine	飢	기	饑
Ji'an	City	吉安	지안	吉安
Jiangsu	Province	江蘇	장d	江蘇
Jiangxi	Province	江西	강서	江西
Jiaxing	City	嘉興	가흥	嘉興
Jie	People	羯	갈	羯
Jilin	Province	吉林	지린	吉林
Jin	Dynasty	金	금	金
Jinhan	People	辰韓	진한	辰韓
Jinju	City	晋州	진주	晉州
Jogan	Era	貞観	조간	貞觀
Jomon	Culture	縄文	조몬	繩文
Joseon	Dynasty	朝鮮	조선	朝鮮
Joseon Minjujueui Inmin Gonghwaguk	Nation	朝鮮民主主義人民共和国	조선민주주의인민공화국	朝鮮民主主義人民共和國
Jun	King	準王	준왕	准王
Kabutoyama	Village	甲山	가부토야마	甲山
Kaesong	City	開城	개성	開城
Kagawa	Prefecture	香川	가가와	香川
Kagoshima	City, prefecture	鹿児島	가고시마	鹿兒島
Kaifeng	City	開封	카이펑	開封
kainin kakiage-cho	Literature	懐妊書上帳	회임서상장	懷妊書上帳
kako-cho	Literature	過去帳	과거장	過去帳
Kamakura	Era	鎌倉	가마쿠라	鐮倉
Kan'ei	Era	寛永	칸에이	寬永
Kanagawa	Prefecture	神奈川	가나가와	神奈川
Kangxi	Emperor	康熙帝	강희제	康熙帝
Kanki	Era	寛喜	간기	寬喜
kanoe-inu	Year	かのえいぬ	경술	庚戌
kanoto-i	Year	かのとい	신해	辛亥
Kansho	Era	長禄	조로쿠	長祿
Kanto	Region	関東	간토	關東
Kaohsiung	City, province	高雄	가오슝	高雄
Karenko	Province	花蓮港	화롄항	花蓮港
Kawasaki	City	川崎	가와사키	川崎
Keeling	City	基隆	지룽	基隆

(continued)

(continued)

English	Category	Japanese	Korean	Chinese
Khitan	People	契丹	거란	契丹
Kinai	Region	畿内	기내	畿内
Kinmen	Islands	金門	진먼	金門
kinoto-i	Year	きのとい	乙亥	乙亥
Kobe	City	神戸	고베	神戸
Kochi	City, prefecture	高知	고치	高知
Kofun	Culture	古墳	고분	古墳
Koxinga	Person	鄭成功	정성공	鄭成功
Kumamoto	City, prefecture	熊本	구마모토	熊本
Kyoho	Era	享保	교호	享保
Kyoto	City, prefecture	京都	교토	京都
Kyushu	Region	九州	규슈	九州
Later Jin	Kingdom	後晋	후진	後晉
Lee Myung-Bak	Person	李明博	이명박	李明博
Lelang	Commandery	楽浪	낙랑	樂浪
Li Yuan	Person	李淵	이연	李淵
Li Zicheng	Person	李自成	이자성	李自成
Liao	Dynasty	遼	요	遼
Liaodong	Region	遼東	랴오둥	遼東
Liaoning	Province	遼寧	랴오닝	遼寧
Lienchiang	County	連江	롄장	連江
Liu Song	State	劉宋	유송	劉宋
Longshan	Culture	龍山	룽산	龍山
Luoyang	City	洛陽	낙양	洛陽
Mahan	People	馬韓	마한	馬韓
Manchu	Region, people	満洲	만주	滿洲
Manchukuo	State	満洲国	만주국	滿洲國
Mao Zedong	Person	毛沢東	모택동	毛澤東
Masan	City	馬山	마산	馬山
Meiji	Era	明治	메이지	明治
Meireki	Era	明暦	메이레키	明曆
Miaoli	County	苗栗	먀오리	苗栗
Mie	Prefecture	三重	미에	三重
Min	Region, language	閩	민	閩
Min	Queen	閔妃	민비	閔妃
Minamoto	Clan	源	미나모토	源

(continued)

(continued)

English	Category	Japanese	Korean	Chinese
Minamoto no Yoritomo	Person	源頼朝	미나모토노 요리토모	源ɪ朝
Ming	Dynasty	明	명	明
Minnan	Language	閩南	민남	閩南
Miyagi	Prefecture	宮城	미야기	宮城
Miyazaki	City, prefecture	宮崎	미야자키	宮崎
Moon Jae-In	Person	文在寅	문재인	文在寅
Mt. Asama	Mountain	浅間山	아사마야마	淺間山
Mt. Fuji	Mountain	富士山	후지산	富士山
Mt. Paektu	Mountain	白頭山	백두산	長白山
Mt. Tarumae	Mountain	樽前山	다루마에산	樽前山
Mt. Unzen	Mountain	雲仙岳	운젠다케	雲仙岳
Muromachi	Era	室町	무로마치	室町
Myeon	Village	面	면	面
Nagano	City, prefecture	長野	나가노	長野
Nagasaki	City, prefecture	長崎	나가사키	長崎
Nanjing	City	南京	난징	南京
Nantou	County	南投	난터우	南投
Nara	City, prefecture	奈良	나라	奈良
New Taipei	City	新北	신베이	新北
Nian Rebellion	Rebellion	捻軍起義	염군의난	捻亂
Nihon Shoki	Literature	日本書紀	일본서기	日本書紀
Niigata	City, prefecture	新潟	니가타	新瀉
Niita	Village	仁井田	니이다	仁井田
Ningxia	Province	寧夏	닝샤	寧夏
Nishijo	Village	西條	니시조	西條
Nishinomiya	City	西宮	니시노미야	西宮
nobi	Class	奴婢	노비	奴婢
Nobi	Region	濃尾	노비	濃尾
Nojimazaki	Place	野島崎	노지마자키	野島岬
Nomo	Village	野茂	노모	野茂
Northern Song	Dynasty	北宋	북송	北宋
Northern Wei	State	北魏	북위	北魏
Oda Nobunaga	Person	織田信長	오다 노부나가	織田信長
Odawara	City	小田原	오다와라	小田原
Oei	Era	応永	오에이	應永
Oita	City, prefecture	大分	오이타	大分

(continued)

(continued)

English	Category	Japanese	Korean	Chinese
Okayama	City, prefecture	岡山	오카야마	岡山
Okinawa	Prefecture	沖縄	오키나와	沖縄
Okjeo	People	沃沮	옥저	沃沮
on-yeok	Epidemic	瘟疫	온역	瘟疫
Onin	Era	応仁	오닌	應仁
Osaka	City, prefecture	大阪	오사카	大阪
Otabumi	Literature	大田文	오타부미	大田文
Ou	Region	奥羽	오우	奥羽
paoja	Organization	保甲	보갑	保甲
Park Chung-Hee	Person	朴正熙	박정희	朴正熙
Park Geun-Hye	Person	朴槿恵	박근혜	朴槿惠
Penghu	Islands, county	膨湖	팽호	膨湖
Peyongan	Province	平安	평안	平安
Peyongnam	Province	平南	평남	平南
Pingtung	County	屏東	핑둥	屏東
Pohang	City	浦項	포항	浦項
Pyeongbuk	Province	平北	평북	平北
Pyongyang	City	平壌	평양	平壤
Qiang	People	羌	창	羌
Qin	Dynasty	秦	진	秦
Qin Shi Huang	Emperor	秦始皇	진시황	秦始皇
Qing	Dynasty	清	청	清
Qingdao	City	青島	칭다오	青島
Qinghai	Province	青海	칭하이	青海
Qinzong	Emperor	欽宗	흠종	欽宗
Quanlong	Emperor	乾隆帝	건륭제	乾隆帝
Rason	City	羅先	라선	羅先
Rehe	Province	熱河	러허	熱河
Reiwa	Era	令和	레이와	令和
Roh Moo-Hyun	Person	盧武鉉	노무현	盧武鉉
Roh Tae-Woo	Person	盧泰愚	노태우	盧泰愚
Ryanggang	Province	両江	량강	兩江
Ryukyu	Islands	琉球	류큐	琉球
Saga	City, prefecture	佐賀	사가	佐賀
Saitama	Prefecture	埼玉	사이타마	崎玉
Samguk Sagi	Literature	三国史記	삼국사기	三國史記
Samguk Yusa	Literature	三国遺事	삼국유사	三國遺事

(continued)

(continued)

English	Category	Japanese	Korean	Chinese
San'in	Region	山陰	산인	山陰
sankin kotai	Regulation	参勤交代	참근교대	參勤交代
Sanriku	Region	三陸	산리쿠	三陸
Sanyo	Region	山陽	산요	山陽
Sejong	City	世宗	세종	世宗
Sendai	City	仙台	센다이	仙台
Seongnam	City	城南	성남	城南
Seoul	City	ソウル	서울	首爾
Seto Inland	Place	瀬戸内	세토우치	(戶内
Shaanxi	Province	陝西	산시	陝西
Shandong	Province	山東	산둥	山東
Shang	Dynasty	商	상	商
Shang Yang	Person	商鞅	상앙	商鞅
Shanghai	City	上海	상해	上海
Shanxi	Province	山西	산시	山西
Shenyang	City	瀋陽	심양	瀋陽
Shenzhen	City	深圳	선전	深圳
Shi	Urban	市	시	市
Shiga	Prefecture	滋賀	시가	滋賀
Shiji	Literature	史記	사기	史記
Shikoku	Region	四国	시코쿠	四國
Shimabara Rebellion	War	島原の乱	시마바라의난	島原之亂
Shimane	Prefecture	島根	시마네	島根
Shimomoriya	Village	下守屋	시모모리야	下守屋
shin-hae	Year	かのとい	신해	辛亥
Shinsen Shojiroku	Literature	新撰姓氏録	신찬성씨록	新撰姓氏錄
Shirakawa	Village	白河	시라카와	白河
Shizuoka	City, prefecture	静岡	시즈오카	靜岡
Shoka	Era	正嘉	쇼카	正嘉
Showa	Era	昭和	쇼와	昭和
Shu Han	State	蜀漢	촉한	蜀漢
Shugaisho	Literature	拾芥抄	슈가이쇼	拾芥抄
shumon aratame-cho	Literature	宗門改張	종문개장	宗門改張
Shunzi	Emperor	順治帝	준지제	順治帝
Sichuan	Province	四川	사천	四川
Silla	Kingdom	新羅	신라	新羅
Sima Qian	Person	司馬遷	사마천	司馬遷

(continued)

(continued)

English	Category	Japanese	Korean	Chinese
Socheonri	Village	召村里	소촌리	召村里
Song	Dynasty	宋	송	宋
sonno joi	Movement	尊王攘夷	존황양이	尊王攘夷
Southern Chen	State	陳	진	陳
Southern Song	Dynasty	南宋	남송	南宋
Sui	Dynasty	隋	수	隋
Suiyuan	Province	綏遠	d이위안	綏遠
Sujin	Emperor	崇神	스진	崇神
Suwa	County	諏訪	스와	諏訪
Suwon	City	水原	수원	水原
Suzhou	City	蘇州	d저우	蘇州
Taedong	River	大同	대동	大同
Taichung	City, county	台中	타이중	台中
Tainan	City, county	台南	타이난	台南
Taipei	City, county	台北	타이베이	台北
Taiping Rebellion	Rebellion	太平天国の乱	태평천국의란	太平天國之亂
Taira	Clan	平	다이라	平
Taisho	Era	大正	다이쇼	大正
Taitung	City, county	台東	타이둥	台東
Takayama	Town	高山	다카야마	高山
Tanaka Kakuei	Person	田中角栄	타나카 카쿠에이	田中角榮
Tang	Dynasty	唐	당	唐
Taoyuan	City	桃園	타오위안	桃園
Tenmei	Era	天明	덴메이	拂曉
Tenmon	Era	天文	덴분	天文
Tenpo	Era	天保	덴포	天保
Tianjin	City	天津	텐진	天津
Tochigi	Prefecture	栃木	도치기	櫪木
Tohoku	Region	東北	도호쿠	東北
Tokai	Region	東海	도카이	東海
Tokugawa Ieyasu	Person	徳川家康	도구가와 이에야스	德川家康
Tokushima	City, prefecture	徳島	도쿠시마	德島
Tokyo	City, prefecture	東京	도쿄	東京
tondenhei	Soldiers	屯田兵	둔전병	屯田兵
Tongzhi	Literature	通志	통지	通志
Tottori	City, prefecture	鳥取	돗토리	鳥取

(continued)

(continued)

English	Category	Japanese	Korean	Chinese
Toyama	City, prefecture	富山	도야마	富山
Toyo	Queen	台与	도요	臺與
Toyotomi Hideyoshi	Person	豊臣秀吉	도요토미 히데요시	豐臣秀吉
Tozan	Region	東山	도산	東山
Tsushima	Island	対馬	N시마	對馬
Tungning	Kingdom	東寧	동녕	東寧
Ugeo	King	右渠	우거	右渠
Ulsan	City	蔚山	울산	蔚山
Un'yo	Gunboat	雲鷹	운요	雲鷹
Wa Japanese	People	倭人	왜인	倭人
Wakayama	City, prefecture	和歌山	와카야마	和歌山
Wamyo Ruijusho	Literature	和名類聚抄	화명류취초	和名類聚抄
Wang Geon	Person	王建	왕건	王建
wei-fu xian-lao	Early aging	未富先老	미부선로	未富先老
wen-yi	Epidemic	瘟疫	온역	瘟疫
Western Jin	State	西晋	서진	西晉
Western Xia	State	西夏	서하	西夏
Wiman	Person	衛満	위만	衛滿
Wonsan	City	元山	원산	元山
Wu	Language	呉	오	吴
Wu	Emperor	武帝	무제	武帝
Wu Guang	Person	呉広	오광	吳廣
Xi'an	City	西安	시안	西安
Xia	Dynasty	夏	하	夏
Xiajiadian	Culture	夏家店	샤자뎬	夏家店
Xian	County	県	현	縣
Xianbei	People	鮮卑	선비	鮮卑
Xiangcun	Village	郷	향	鄉
Xijing	City	西京	시징	西京
Xikang	Province	西康	시캉	西康
xin-hai	Year	かのとい	신해	辛亥
Xinjiang	Region	新疆	신장	新疆
Xinle Xianzhi	Literature	新楽県志	신락현사	新樂縣志
Xiongnu	People	匈奴	흉노	匈奴
Xuanzong	Emperor	玄宗	현종	玄宗
Yamagata	City, prefecture	山形	야마가타	山形

(continued)

(continued)

English	Category	Japanese	Korean	Chinese
Yamaguchi	City, prefecture	山口	야마구치	山口
Yamanashi	Prefecture	山梨	야마나시	山梨
Yamatai	Kingdom	邪馬台国	야마타이코쿠	邪馬台國
Yamato Japanse	People	大和民族	야마토민족	大和民族
Yan	Kingdom	燕	연	燕
Yan Guifei	Person	楊貴妃	양귀비	楊貴妃
Yang	Emperor	煬帝	양제	煬帝
Yang Seong-Ji	Person	梁誠之	양성지	梁誠之
Yang Xuangan	Person	楊玄感	양현감	楊玄感
yangban	Class	両班	양반	兩班
Yayoi	Culture	弥生	야요이	彌生
Ye-Maek	People	濊貊	예맥	濊貊
yeo	Epidemic	癘	여	癘
yeok	Epidemic	疫	역	疫
Yeongi	County	燕岐	연기	燕岐
yeoyeok	Epidemic	癘疫	여역	癘疫
yi	Epidemic	疫	역	疫
Yi Joseon	Dynasty	李氏朝鮮	이씨조선	李氏朝鮮
Yi Seong-gye	Person	李成桂	이성계	李成桂
yi-hai	Year	きのとい	乙亥	乙亥
Yilan	county	宜蘭	이란	宜蘭
Yokohama	City	横浜	요코하마	横濱
Yokouchi	Village	横内	요코우치	横内
Yongzheng	Emperor	雍正帝	옹정제	雍正帝
You	King	幽王	유왕	幽王
Yowa	Era	養和	요와	養和
Yuan	Dynasty	元	원	元
Yuan Shikai	Person	袁世凱	위안스카이	袁世凱
Yue	Language	粵	웨	粵
Yunlin	County	雲林	윈린	雲林
Yunnan	Province	雲南	윈난	雲南
Zenkoji	Temple	善光寺	젠코지	善光寺
zenpokoenfun	Culture	前方後円墳	전방후원분	前方後円墳
zhang-yi	Epidemic	瘴疫	장역	瘴疫
Zhao Gou	Person	趙構	조구	趙構
Zhao Kuangyin	Person	趙匡胤	조광윤	趙匡胤
Zhejiang	Province	浙江	절강	浙江

(continued)

(continued)

English	Category	Japanese	Korean	Chinese
Zhen	Town	鎮	진	鎮
Zheng	King	政	정	政
Zhengzhou	City	鄭州	정저우	鄭州
Zhenzong	Emperor	真宗	진종	真宗
Zhili	Province	直隷	즈리	直隷
Zhixiaoshi	Metropolis	直轄市	직할시	直轄市
Zhou	Dynasty	周	주	周
Zhu Xi	Person	朱熹	주희	朱熹
Zhu Yuanzhang	Person	朱元璋	주원장	朱元璋

Maps

Because there was no adequate set of four blank maps, I had to draw maps by myself. I intended the minimal drawing to display the positioning of administrative areas. Thus, small islands were omitted, and adjacent countries were not shown. Although not specified in the legend, prefectures of Japan are classified into four classes: a circuit (Hokkaido), a metropolis (Tokyo), urban prefectures (Osaka and Kyoto), and other ordinary prefectures. For other countries, classes were clarified by adding "City," "Province," "County," or "Autonomous Region." (Maps 1, 2, 3 and 4).

T. Suzuki, *Comparative Population History of Eastern Asia*,
https://doi.org/10.1007/978-981-99-9367-3

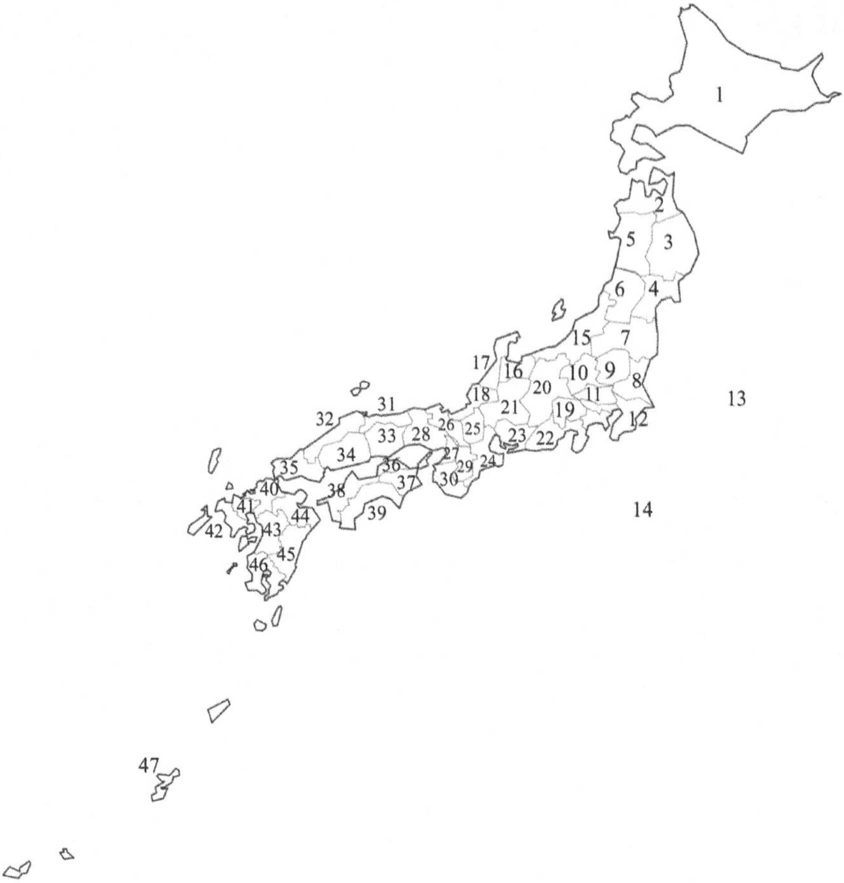

1 Hokkaido	13 Tokyo	25 Shiga	37 Kagawa
2 Aomori	14 Kanagawa	26 Kyoto	38 Ehime
3 Iwate	15 Niigata	27 Osaka	39 Kochi
4 Miyagi	16 Toyama	28 Hyogo	40 Fukuoka
5 Akita	17 Ishikawa	29 Nara	41 Saga
6 Yamagata	18 Fukui	30 Wakayama	42 Nagasaki
7 Fukushima	19 Yamanashi	31 Tottori	43 Kumamoto
8 Ibaraki	20 Nagano	32 Shimane	44 Oita
9 Tochigi	21 Gifu	33 Okayama	45 Miyazaki
10 Gunma	22 Shizuoka	34 Hiroshima	46 Kagoshima
11 Saitama	23 Aichi	35 Yamaguchi	47 Okinawa
12 Chiba	24 Mie	36 Tokushima	

Map 1 Prefectures of Japan

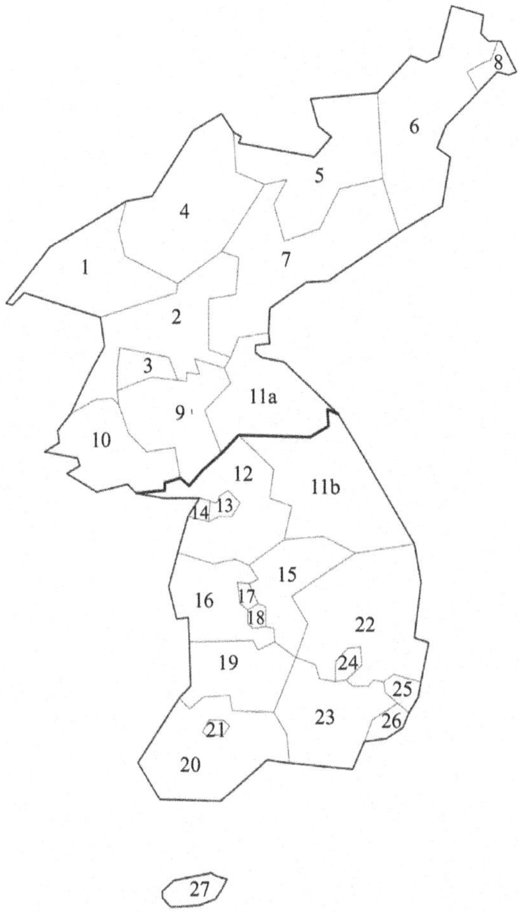

1 Pyeongbuk Province
2 Pyeongnam Province
3 Pyongyang City
4 Chagang Province
5 Ryanggang Province
6 Hambuk Province
7 Hamnam Province
8 Rason City
9 Hwangbuk Province
10 Hwangnam Province
11a Gangweon Province (north)
11b Gangweon Province (south)
12 Gyeonggi Province
13 Seoul City
14 Incheon City
15 Chungbuk Province
16 Chungnam Province
17 Sejong City
18 Daejeon City
19 Jeonbuk Province
20 Jeonnam Province
21 Gwangju City
22 Gyeongbuk Province
23 Gyeongnam Province
24 Daegu City
25 Ulsan City
26 Busan City
27 Jeju Province

Map 2 Major cities and provinces of Korea

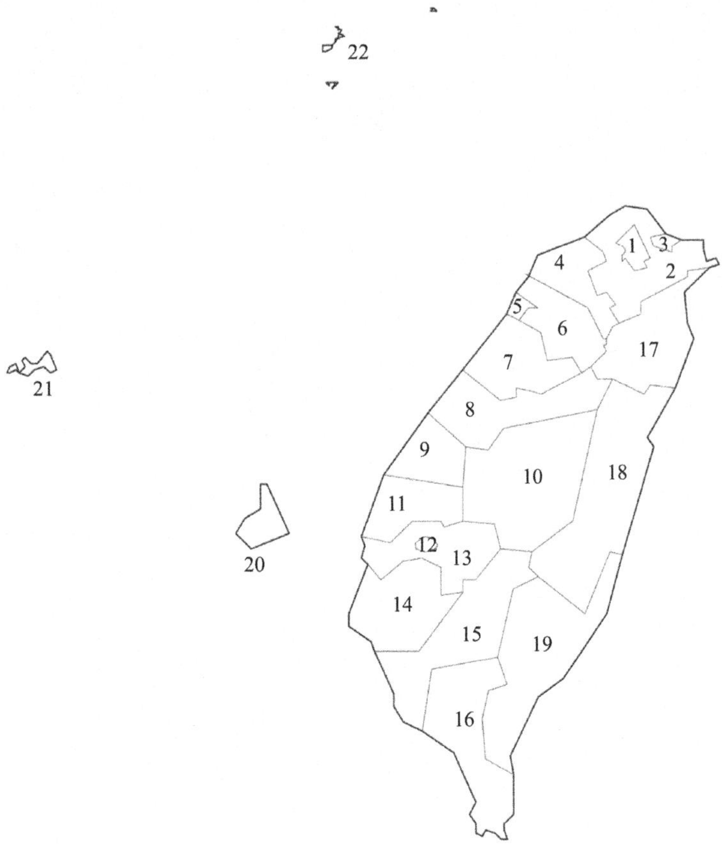

1 Taipei City
2 New Taipei City
3 Keelung City
4 Taoyuan City
5 Hsinchu City
6 Hsinchu County
7 Miaoli County
8 Taichung City
9 Changhua County
10 Nantou County
11 Yunlin County
12 Chiayi City
13 Chiayi County
14 Tainan City
15 Kaohsiung City
16 Pingtung County
17 Yilan County
18 Hualien County
19 Taitung County
20 Penghu County
21 Kinmen County
22 Lienchiang County

Map 3 Major cities and countries of Taiwan

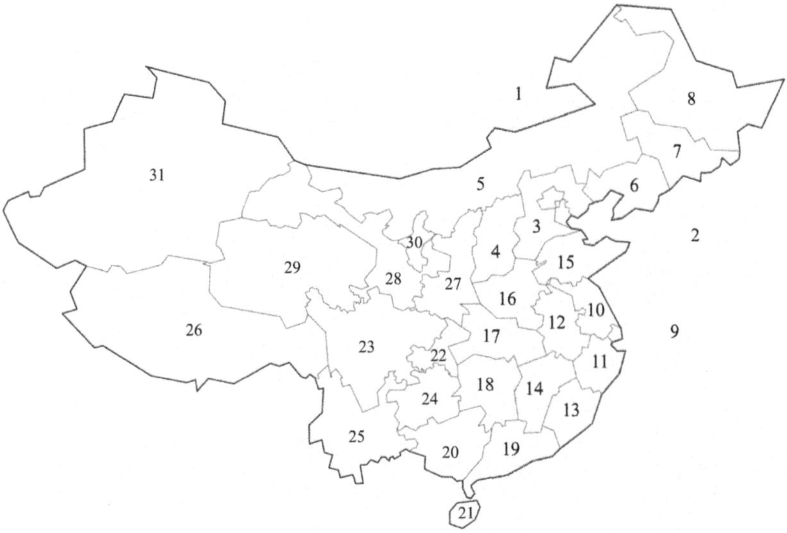

1 Beijing City
2 Tianjin City
3 Hebei Province
4 Shanxi Province
5 Inner Mongolia
6 Liaoning Province
7 Jilin Province
8 Heilongjiang Province
9 Shanghai City
10 Jiangsu Province
11 Zhejiang Province
12 Anhui Province
13 Fujian Province
14 Jiangxi Province
15 Shandong Province
16 Henan Province
17 Hubei Province
18 Hunan Province
19 Guangdong Province
20 Guangxi Zhuang Autonomous Region
21 Hainan Province
22 Chongqing City
23 Sichuan Province
24 Guizhou Province
25 Yunnan Province
26 Tibet Autonomous Region
27 Shaanxi Province
28 Gansu Province
29 Qinghai Province
30 Ningxia Hui Autonomous Region
31 Xinjiang Uygur Autonomous Region

Map 4 Major cities and provinces of China

SPRINGER NATURE

GPSR Compliance

The European Union's (EU) General Product Safety Regulation (GPSR) is a set of rules that requires consumer products to be safe and our obligations to ensure this.

If you have any concerns about our products, you can contact us on ProductSafety@springernature.com

In case Publisher is established outside the EU, the EU authorized representative is:

Springer Nature Customer Service Center GmbH
Europaplatz 3
69115 Heidelberg, Germany

The manufacturer's authorised representative in the EU is Springer Nature Customer Service Centre GmbH, Europaplatz 3, 69115 Heidelberg, Germany. If you have any concerns regarding our products, please contact ProductSafety@springernature.com

Printed and bound by CPI Group (UK) Ltd, Croydon, CR0 4YY
30/09/2026
02225318-0003